TECHNOLOGY, MANAGEMENT & SOCIETY

*the text of this book is printed
on 100% recycled paper*

Books by Peter F. Drucker

TECHNOLOGY, MANAGEMENT & SOCIETY

Essays by Peter F. Drucker

HARPER COLOPHON BOOKS
Harper & Row, Publishers
New York, Hagerstown, San Francisco, London

Major portions of this work were previously published in *Harvard Business Review*, *Technology and Culture*, *Management Today*, *The Journal of Business of the University of Chicago* and *The McKinsey Quarterly*.

Chapters 4 and 5 are reprinted from *Technology in Western Civilization*, Volume II, edited by Melvin Kranzberg and Carroll W. Pursell, Jr. Copyright © 1967 by the Regents of the University of Wisconsin. Reprinted by permission of Oxford University Press.

First HARPER COLOPHON edition published 1977
ISBN: 0–06–090569–7

Contents

Preface

There should be underlying unity to a collection of essays. There should be a point of view, a central theme, an organ point around which the whole volume composes itself. And there is, I believe, such fundamental unity to this volume of essays, even though they date from more than a dozen years ago and discuss a variety of topics. One of the essays, "Work and Tools," states: "Technology is not about tools, it deals with how Man works." This might be the device of this entire volume, if not, indeed, for my entire work over the years.

All the essays in this volume deal with one or the other aspect of what used to be called "the material civilization": they all deal with man's tools and his materials, with his institutions and organizations, and with the way he works and makes his living. But throughout, work and materials, organizations and a living are seen as "extensions of man," rather than as material artifacts and part of inanimate nature. If I were to reflect on my own position over the years, I would say that, from the very beginning, I rejected the common nineteenth-century view which divided man's society into "culture," dealing with ideas and symbols, and "civilization," dealing with artifacts and things. "Civilization" to me has always been a part of man's personality, and an area in which he expressed his basic ideals, his dreams, his aspirations, and his values. Some of the essays in this volume are about technology and its history. Some are about management and managers. Some are about specific tools—the computer, for instance. But all of them are about man at work; all are about man trying to make himself effective.

An essay collection, however, should also have diversity. It should break an author's thought and work the way a prism breaks light. Indeed, the truly enjoyable essay collection is full of surprises as the same author, dealing with very much the same areas, is suddenly revealed in new guises and suddenly reveals new facets of his subject. The essays collected in this volume deal with only one of the major areas that have been of concern to me—the area of the "material civilization." But there is a good deal of variety in them. Five of the twelve essays in this volume deal with technology, its history and its impact on man and his culture. They range in time, however, from a look at the "first technological revolution," seven thousand years ago, when the irrigation cities created what we still call "modern civilization," to an attempt to evaluate the position of technology in our present century. They all assume that history cannot be written, let alone make sense, unless it takes technology into account and is aware of the development of man's tools and his use of them through the ages. This, needless to say, is not a position historians traditionally have held; there are only signs so far that they are beginning to realize that technology has been with us from the earliest date and has always been an intimate and integral part of man's experience, man's society, and man's history. At the same time, these essays all assume that the technologist, to use his tools constructively, has to know a good deal of history and has to see himself and his discipline in relationship to man and society—and that has been an even less popular position among technologists than the emphasis on technology has been among historians.

Four essays in this volume—the first two, the essay, "The Once and Future Manager," and the essay on "Business Objectives and Survival Needs"—look upon the manager as the agent of today's society and upon management as a central social function. They assume that managers handle tools, assume that managers know their tools thoroughly and are willing to acquire new ones as needed. But, above all, they ask the question, "What *results* do we expect from the manager; what *results* does his enterprise, whether a business or a government agency, need from him? What results, above all, do our society and the human beings that compose it have a right to expect from a manager and from management?" The concern is with management as it affects the quality of life—that management can provide the quantities of life is taken as proven.

The remaining three essays ("Long-Range Planning," "The Manager and the Moron," and "Can Management Ever Be a Science?") deal with basic approaches and techniques. They are focused on management within the enterprise rather than on management as a social function. But they stress constantly the purpose of management, which is not to be efficient but to be productive, for the human being, for economy, for society.

An essay collection, finally, should convey the personality of the author better than a book can. This is why I enjoy reading essays. It should bring out a man's style, a man's wit, and the texture of a man's mind. Whether this essay collection does this, I leave to the reader to judge. But I do hope that these twelve essays of mine, written for different purposes and at different times over the last twelve years, will also help to establish the bond between author and writer, which, in the last analysis, is why a writer writes and a reader reads.

PETER F. DRUCKER

Montclair, New Jersey
New Year's Day, 1970

TECHNOLOGY, MANAGEMENT & SOCIETY

1 | Information, Communications and Understanding

Concern with "information" and "communications" started shortly before World War I. Russell and Whitehead's *Principia Mathematica*, which appeared in 1910, is still one of the foundation books. And a long line of illustrious successors—from Ludwig Wittgenstein through Norbert Wiener and A. N. Chomsky's "mathematical linguistics" today—has continued the work on the *logic* of information. Roughly contemporaneous is the interest in the *meaning* of communication; Alfred Korzybski started on the study of "general semantics," i.e., on the meaning of communications, around the turn of the century. It was World War I, however, which made the entire Western world communications-conscious. When the diplomatic documents of 1914 in the German and Russian archives were published, soon after the end of the fighting, it became appallingly clear that the catastrophe had been caused, in large measure, by communications failure despite copious and reliable information. And the war itself—especially the total failure of its one and only strategic concept, Winston Churchill's Gallipoli campaign in

Paper read before the Fellows of the International Academy of Management, Tokyo, Japan, October, 1969.

1915/16—was patently a tragicomedy of noncommunications. At the same time, the period immediately following World War I—a period of industrial strife and of total noncommunication between Westerners and "revolutionary" Communists (and a little later, equally revolutionary Fascists)—showed both the need for, and the lack of, a valid theory or a functioning practice of communications, inside existing institutions, inside existing societies, and between various leadership groups and their various "publics."

As a result, communications suddenly became, forty to fifty years ago, a consuming interest of scholars as well as of practitioners. Above all, communications in management has this last half century been a central concern to students and practitioners in all institutions—business, the military, public administration, hospital administration, university administration, and research administration. In no other area have intelligent men and women worked harder or with greater dedication than psychologists, human relations experts, managers, and management students have worked on improving communications in our major institutions.

We have more attempts at communications today, that is, more attempts to talk to others, and a surfeit of communications media, unimaginable to the men who, around the time of World War I, started to work on the problems of communicating. The trickle of books on communications has become a raging torrent. I recently received a bibliography prepared for a graduate seminar on communications; it ran to ninety-seven pages. A recent anthology (*The Human Dialogue,* edited by Floyd W. Matson and Ashley Montagu, The Free Press of Glencoe, 1967) contains articles by forty-nine different contributors.

Yet communications has proven as elusive as the unicorn. Each of the forty-nine contributors to *The Human Dialogue* has a theory of communications which is incompatible with all the others. The noise level has gone up so fast that no one

can really listen any more to all that babble about commu
cations. But there is clearly less and less communicati.
The communications gap within institutions and between
groups in society has been widening steadily—to the point
where it threatens to become an unbridgeable gulf of total
misunderstanding.

In the meantime, there is an information explosion. Every
professional and every executive—in fact, everyone except
the deaf-mute—suddenly has access to data in inexhaustible
abundance. All of us feel—and overeat—very much like the
little boy who has been left alone in the candy store. But
what has to be done to make this cornucopia of data redound to information, let alone to knowledge? We get a
great many answers. But the one thing clear so far is that no
one really has an answer. Despite "information theory" and
"data processing," no one yet has actually seen, let alone
used, an "information system," or a "data base." The one
thing we do know, though, is that the abundance of information changes the communications problem and makes it both
more urgent and even less tractable.

There is a tendency today to give up on communications.
In psychology, for instance, the fashion today is the T-group
with its "sensitivity training." The avowed aim is not communications, but self-awareness. T-groups focus on the "I"
and not on the "thou." Ten or twenty years ago the rhetoric
stressed "empathy"; now it stresses "doing one's thing." However needed self-knowledge may be, communication is
needed at least as much (if, indeed, self-knowledge is possible without action on others, that is, without communications). Whether the T-groups are sound psychology and
effective psychotherapy is well beyond my competence and
the scope of this paper. But their popularity attests to the
failure of our attempts at communications.

Despite the sorry state of communications in theory and
practice, we have, however, learned a good deal about in-

formation and communications. Most of it, though, has not come out of the work on communications to which we have devoted so much time and energy. It has been the by-product of work in a large number of seemingly unrelated fields, from learning theory to genetics and electronic engineering. We equally have a lot of experience—though mostly of failure—in a good many practical situations in all kinds of institutions. Communications we may, indeed, never understand. But communications in organizations— call it *managerial communications*—we do know something about by now. It is a much narrower topic than communications per se—but it is the topic to which this paper shall address itself.

We are, to be sure, still far away from mastery of communications, even in organizations. What knowledge we have about communications is scattered and, as a rule, not accessible, let alone in applicable form. But at least we increasingly know what does not work and, sometimes, why it does not work. Indeed, we can say bluntly that most of today's brave attempts at communication in organizations— whether business, labor unions, government agencies, or universities—is based on assumptions that have been proven to be invalid—and that, therefore, these efforts cannot have results. And perhaps we can even anticipate what might work.

What We Have Learned

We have learned, mostly through doing the wrong things, the following four fundamentals of communications:

(1) Communication is perception,
(2) Communication is expectations,
(3) Communication is involvement,

(4) Communication and information are totally different. But information presupposes functioning communications.

Communication Is Perception

An old riddle asked by the mystics of many religions—the Zen Buddhists, the Sufis of Islam, or the rabbis of the Talmud—asks: "Is there a sound in the forest if a tree crashes down and no one is around to hear it?" We now know that the right answer to this is "no." There are sound waves. But there is no sound unless someone perceives it. Sound is created by perception. Sound is communication.

This may seem trite; after all, the mystics of old already knew this, for they, too, always answered that there is no sound unless someone can hear it. Yet the implications of this rather trite statement are great indeed.

(a) First, it means that it is the recipient who communicates. The so-called communicator, that is, the person who emits the communication, does not communicate. He utters. Unless there is someone who hears, there is no communication. There is only noise. The communicator speaks or writes or sings—but he does not communicate. Indeed he cannot communicate. He can only make it possible, or impossible, for a recipient—or rather percipient—to perceive.

(b) Perception, we know, is not logic. It is experience. This means, in the first place, that one always perceives a configuration. One cannot perceive single specifics. They are always part of a total picture. *The Silent Language* (as Edward T. Hall called it in the title of his pioneering work ten years ago)—that is, the gestures, the tone of voice, the environment all together, not to mention the cultural and social referents—cannot be dissociated from the spoken language. In fact, without them the spoken word has no meaning and cannot communicate. It is not only that the same words, e.g., "I enjoyed meeting you," will be heard as having a wide variety of meanings. Whether they are heard as warm or as icy cold, as endearment or as rejection, depends on their setting in the silent language, such as the tone of voice or the occasion. More important is that by themselves, that is, without being part of the total configuration of occasion, value, silent language, and so on, the phrase has no meaning at all. By itself it cannot make possible communication. It cannot be understood. Indeed, it cannot be heard. To paraphrase an old proverb of the Human Relations school: "One cannot communicate a word; the whole man always comes with it."

(c) But we know about perception also that one can only perceive what one is capable of perceiving. Just as the human ear does not hear sounds above a certain pitch, so does human perception all together not perceive what is beyond its range of perception. It may, of course, hear physically, or see visually, but it cannot accept. The stimulus cannot become communication.

This is a very fancy way of stating something the teachers of rhetoric have known for a very long time—though the practitioners of communications tend to forget it again and again. In Plato's *Phaedrus*, which, among other things, is also the earliest extant treatise on rhetoric, Socrates points out that one has to talk to people in terms of their own experience, that is, that one has to use a carpenter's metaphors when talking to carpenters, and so on. One can only communicate in the recipient's language or altogether in his terms. And the terms have to be experience-based. It, therefore, does very little good to try to explain terms to people. They will not be able to receive them if the terms are not of their own experience. They simply exceed their perception capacity.

The connection between experience, perception, and concept formation, that is, cognition, is, we now know, infinitely subtler and richer than any earlier philosopher imagined. But one fact is proven and comes out strongly in the most disparate work, e.g., that of Piaget in Switzerland, that of B. F. Skinner of Harvard, or that of Jerome Bruner (also of Harvard). Percept and concept in the learner, whether child or adult, are not separate. We cannot perceive unless we also conceive. But we also cannot form concepts unless we can perceive. To communicate a concept is impossible unless the recipient can perceive it, that is, unless it is within his perception.

There is a very old saying among writers: "Difficulties with a sentence always mean confused thinking. It is not the sentence that needs straightening out, it is the thought behind it." In writing we attempt, of course, to communicate with ourselves. An unclear sentence is one that exceeds our own capacity for perception. Working on the sentence, that is, working on what is normally called communications, cannot solve the problem. We have to work on our own concepts first to be able to understand what we are trying to say—and only then can we write the sentence.

In communicating, whatever the medium, the first question has to be, "Is this communication within the recipient's range of perception? Can he receive it?"

The "range of perception" is, of course, physiological and largely (though not entirely) set by physical limitations of man's animal body. When we speak of communications, however, the most important limitations on perception are usually cultural and emotional rather than physical. That fanatics are not being convinced by rational arguments, we have known for thousands of years. Now we are beginning to understand that it is not "argument" that is lacking. Fanatics do not have the ability to perceive a communication which goes beyond their range of emotions. Before this is possible, their emotions would have to be altered. In other words, no one is really "in touch with reality," if by that we mean complete openness to evidence. The distinction between sanity and paranoia is not in the ability to perceive, but in the ability to learn, that is, in the ability to change one's emotions on the basis of experience.

That perception is conditioned by what we are capable of perceiving was realized forty years ago by the most quoted but probably least heeded of all students of organization, Mary Parker Follett, especially in her collected essays, *Dynamic Administration* (New York, Harper's, 1941). Follett taught that a disagreement or a conflict is likely not to be about the answers, or, indeed, about anything ostensible. It is, in most cases, the result of incongruity in perceptions. What A sees so vividly, B does not see at all. And, therefore, what A argues has no pertinence to B's concerns, and vice versa. Both, Follett argued, are likely to see reality. But each is likely to see a different aspect thereof. The world, and not only the material world, is multidimensional. Yet one can only see one dimension at a time. One rarely realizes that there could be other dimensions, and that something that is so obvious to us and so clearly validated by our emotional experience has other dimensions, a back and sides, which are entirely different and which, therefore, lead to entirely different perception. The old story about the blind men and the elephant in which every one of them, upon encountering this strange beast, feels one of the elephant's parts, his leg, his trunk, his hide, and reports an entirely different conclusion, each held tenaciously, is simply a story of the human condition. And there is no possibility of communication until this is understood and until he who has felt the hide of the elephant goes over to him who has felt the leg and feels the leg himself. There is no possibility of communications, in other words, unless we first know what the recipient, the true communicator, can see and why.

Communication Is Expectations

We perceive, as a rule, what we expect to perceive. We see largely what we expect to see, and we hear largely what we expect to hear. That the unexpected may be resented is not the important thing—though most of the writers on communications in business or government think it is. What is truly important is that the unexpected is usually not received at all. It is either not seen or heard but ignored. Or it is misunderstood, that is, mis-seen as the expected or misheard as the expected.

On this we now have a century or more of experimentation. The results are quite unambiguous. The human mind attempts to fit impressions and stimuli into a frame of expectations. It resists vigorously any attempts to make it "change its mind," that is, to perceive what it does not expect to perceive or not to perceive what it expects to perceive. It is, of course, possible to alert the human mind to the fact that what it perceives is contrary to its expectations. But this first requires that we understand what it expects to perceive. It then requires that there be an unmistakable signal—"this is different," that is, a shock which breaks continuity. A "gradual" change in which the mind is supposedly led by small, incremental steps to realize that what is perceived is not what it expects to perceive will not work. It will rather reinforce the expectations and will make it even more certain that what will be perceived is what the recipient expects to perceive.

Before we can communicate, we must, therefore, know what the recipient expects to see and to hear. Only then can we know whether communication can utilize his expectations—and what they are—or whether there is need for the "shock of alienation," for an "awakening" that breaks through the recipient's expectations and forces him to realize that the unexpected is happening.

Communication Is Involvement

Many years ago psychologists stumbled on a strange phenomenon in their studies of memory, a phenomenon that, at first, upset all their hypotheses. In order to test memory, the psychologists compiled a list of words to be shown to their experimental subjects for varying times as a test of their retention capacity. As control, a list of nonsense words, mere jumbles of letters, were devised to find out to what extent understanding influenced memory. Much to the surprise of these early experimenters almost a century ago or so, their subjects (mostly students, of course,) showed totally uneven memory retention of individual words. More surprising, they showed amazingly high retention of the nonsense words. The explanation of the first phenomenon is fairly obvious. Words are not mere information. They do carry emotional charges. And, therefore, words with unpleasant or threatening associations tend to be suppressed, words with pleasant associations retained. In fact, this selective retention by emotional association has since been used to construct tests for emotional disorders and for personality profiles.

The relatively high retention rate of nonsense words was a greater problem. It was expected, after all, that no one would really remember words that had no meaning at all. But it has become clear over the years that the memory for these words, though limited, exists precisely because these words have no meaning. For this reason, they also make no demand. They are truly neuter. In respect to them, memory could be said to be truly mechanical, showing neither emotional preference nor emotional rejection.

A similar phenomenon, known to every newspaper editor,

is the amazingly high readership and retention of the fillers, the little three- or five-line bits of irrelevant incidental information that are being used to balance a page. Why should anybody want to read, let alone remember, that it first became fashionable to wear different-colored hose on each leg at the court of some long-forgotten duke? Why should anybody want to read, let alone remember, when and where baking powder was first used? Yet there is no doubt that these little tidbits of irrelevancy are read and, above all, that they are remembered far better than almost anything in the daily paper except the great screaming headlines of the catastrophes. The answer is that these fillers make no demands. It is precisely their total irrelevancy that accounts for their being remembered.

Communications are always propaganda. The emitter always wants "to get something across." Propaganda, we now know, is both a great deal more powerful than the rationalists with their belief in "open discussion" believe, and a great deal less powerful than the mythmakers of propaganda, e.g., a Dr. Goebbels in the Nazi regime, believed and wanted us to believe. Indeed, the danger of total propaganda is not that the propaganda will be believed. The danger is that nothing will be believed and that every communication becomes suspect. In the end, no communication is being received any more. Everything anyone says is considered a demand and is resisted, resented, and in effect not heard at all. The end results of total propaganda are not fanatics, but cynics—but this, of course, may be even greater and more dangerous corruption.

Communication, in other words, always makes demands. It always demands that the recipient become somebody, do something, believe something. It always appeals to motivation. If, in other words, communication fits in with the aspirations, the values, the purposes of the recipient, it is powerful. If it goes against his aspirations, his values, his motivations, it is likely not to be received at all, or, at best, to be resisted. Of course, at its most powerful, communication brings about conversion, that is, a change of personality, of values, beliefs, aspirations. But this is the rare, existential event, and one against which the basic psychological forces of every human being are strongly organized. Even the Lord, the Bible reports, first had to strike Saul blind before he could raise him as Paul. Communications aiming at conversion demand surrender. By and large, therefore, there is no communication unless the message can key in to the recipient's own values, at least to some degree.

Communication and Information Are Different and Largely Opposite—Yet Interdependent

(a) Where communication is perception, information is logic. As such, information is purely formal and has no meaning. It is impersonal rather than interpersonal. The more it can be freed of the human component, that is, of such things as emotions and values, expectations and perceptions, the more valid and reliable does it become. Indeed, it becomes increasingly informative.

All through history, the problem has been how to glean a little information out of communications, that is, out of relationships between people, based on perception. All through history, the problem has been to isolate the information content from an abundance of perception. Now, all of a sudden, we have the capacity to provide information—both because of the conceptual work of the logicians, especially the symbolic logic of Russell and Whitehead, and because of the technical work on data processing and data storage, that is, of course, especially because of the computer and its tremendous capacity to store, manipulate, and transmit data. Now, in other words, we have the opposite problem from the one mankind has always been struggling with. Now we have the problem of handling information per se, devoid of any communication content.

(b) The requirements for effective information are the opposite of those for effective communication. Information is, for instance, always specific. We perceive a configuration in communications; but we convey specific individual data in the information process. Indeed, information is, above all, a principle of economy. The fewer data needed, the better the information. And an overload of information, that is, anything much beyond what is truly needed, leads to a complete information blackout. It does not enrich, but impoverishes.

(c) At the same time, information presupposes communication. Information is always encoded. To be received, let alone to be used, the code must be known and understood by the recipient. This requires prior agreement, that is, some communication. At the very least, the recipient has to know what the data pertain to. Are the figures on a piece of computer tape the height of mountain tops or the cash balances of Federal Reserve member banks? In either case, the recipient would have to know what mountains are or what banks are to get any information out of the data.

The prototype information system may well have been the peculiar language known as *Armee Deutsch* (Army German), which served as language of command in the Imperial Austrian Army prior to 1918. A polyglot army in which officers, noncommissioned officers, and men often had no language in common, it functioned remarkably well with fewer than two hundred specific words, "fire," for instance, or "at ease," each of which had only one totally unambiguous meaning. The meaning was always an action. And the words were learned in and through actions, i.e., in what behaviorists now call operant conditioning. The tensions in the Austrian army after many decades of nationalist turmoil were very great indeed. Social intercourse between members of different nationalities serving in the same unit became increasingly difficult, if not impossible. But to the very end, the information system functioned. It was completely formal, completely rigid, completely logical in that each word had only one possible meaning; and it rested on completely pre-established communication regarding the specific response to a certain set of sound waves. This example, however, shows also that the effectiveness of an information system depends on the willingness and ability to think through carefully what information is needed by whom for what purposes, and then on the systematic creation of communication between the various parties to the system as to the meaning of each specific input and output. The effectiveness, in other words, depends on the pre-establishment of communication.

(d) Communication communicates better the more levels of meaning it has and the less possible it is, therefore, to quantify it.

Medieval esthetics held that a work of art communicates on a number of levels, at least three if not four: the literal, the metaphorical, the allegorical, and the symbolic. The work of art that most consciously converted this theory into artistic practice was, of course, Dante's *Divina Commedia*. If, by information, we mean something that can be quantified, then the *Divina Commedia* is without any information content whatever. But it is precisely the ambiguity, the multiplicity of levels on which this book can be read, from being a fairy tale to being a grand synthesis of metaphysics, that makes it the overpowering work of art it is, and the immediate communication which it has been to generations of readers.

Communications, in other words, may not be dependent on information. Indeed, the most perfect communications may be purely shared experiences, without any logic whatever. Perception has primacy rather than information.

I fully realize that this summary of what we have learned is gross oversimplification. I fully realize that I have glossed over some of the most hotly contested issues in psychology and perception. Indeed, I may well be accused of brushing aside most of the issues which the students of learning and of perception would themselves consider central and important.

But my aim has, of course, not been to survey these big areas. My concern is not with learning or with perception. It is with communications, and, in particular, with communications in the large organization, be it business enterprise, government agency, university, or armed service.

This summary might also be criticized for being trite, if not obvious. No one, it might be said, could possibly be surprised at its statements. They say what everybody knows. But whether this be so or not, it is not what everybody does. On the contrary, the logical implications of these apparently simple and obvious statements for communications in organizations are at odds with current practice and, indeed, deny validity to the honest and serious efforts to communicate which we have been making for many decades now.

What, then, can our knowledge and our experience teach us about communications in organizations, about the reasons for our failures, and about the prerequisites for success in the future?

(1) For centuries we have attempted communication downward. This, however, cannot work, no matter how hard and how intelligently we try. It cannot work, first, because it focuses on what we want to say. It assumes, in other words,

that the utterer communicates. But we know that all he does is utter. Communication is the act of the recipient. What we have been trying to do is to work on the emitter, specifically on the manager, the administrator, the commander, to make him capable of being a better communicator. But all one can communicate downward are commands, that is, prearranged signals. One cannot communicate downward anything connected with understanding, let alone with motivation. This requires communication upward, from those who perceive to those who want to reach their perception.

This does not mean that managers should stop working on clarity in what they say or write. Far from it. But it does mean that how we say something comes only after we have learned what to say. And this cannot be found out by "talking to," no matter how well it is being done. "Letters to the Employees," no matter how well done, will be a waste unless the writer knows what employees can perceive, expect to perceive, and want to do. They are a waste unless they are based on the recipient's rather than the emitter's perception.

(2) But "listening" does not work either. The Human Relations School of Elton Mayo, forty years ago, recognized the failure of the traditional approach to communications. Its answer—especially as developed in Mayo's two famous books, *The Human Problems of an Industrial Civilization* (Boston, Harvard Business School, 1933) and *The Social Problems of an Industrial Civilization* (Boston, Harvard Business School, 1945)—was to enjoin listening. Instead of starting out with what I, that is, the executive, want to get across, the executive should start out by finding out what subordinates want to know, are interested in, are, in other words, receptive to. To this day, the human relations prescription, though rarely practiced, remains the classic formula.

Of course, listening is a prerequisite to communication. But it is not adequate, and it cannot, by itself, work. Perhaps the reason why it is not being used widely, despite the popularity of the slogan, is precisely that, where tried, it has failed to work. Listening first assumes that the superior will understand what he is being told. It assumes, in other words, that the subordinates can communicate. It is hard to see, however, why the subordinate should be able to do what his superior cannot do. In fact, there is no reason for assuming he can. There is no reason, in other words, to believe that listening results any less in misunderstanding and miscommunications than does talking. In addition, the theory of listening does not take into account that communications is involvement. It does not bring out the subordinate's preferences and desires, his values and aspirations. It may explain the reasons for misunderstanding. But it does not lay down a basis for understanding.

This is not to say that listening is wrong, any more than the futility of downward communications furnishes any argument against attempts to write well, to say things clearly and simply, and to speak the language of those whom one addresses rather than one's own jargon. Indeed, the realization that communications have to be upward—or rather that they have to start with the recipient, rather than the emitter, which underlies the concept of listening—is absolutely sound and vital. But listening is only the starting point.

(3) More and better information does not solve the communications problem, does not bridge the communications gap. On the contrary, the more information, the greater is the need for functioning and effective communication. The more information, in other words, the greater is the communications gap likely to be.

The more impersonal and formal the information process in the first place, the more will it depend on prior agreement on meaning and application, that is, on communications. In the second place, the more effective the information process, the more impersonal and formal will it become, the more will it separate human beings and thereby require separate, but also much greater, efforts, to re-establish the human relationship, the relationship of communication. It may be said that the effectiveness of the information process will depend increasingly on our ability to communicate, and that, in the absence of effective communication—that is, in the present situation—the information revolution cannot really produce information. All it can produce is data.

It can also be said—and this may well be more important—that the test of an information system will increasingly be the degree to which it frees human beings from concern with information and allows them to work on communications. The test, in particular, of the computer will be how much time it gives executives and professionals on all levels for direct, personal, face-to-face relationships with other people.

It is fashionable today to measure the utilization of a computer by the number of hours it runs during one day. But this is not even a measurement of the computer's efficiency. It is purely a measurement of input. The only measurement of output is the degree to which availability of information enables human beings not to control, that is, not to spend time trying to get a little information on what happened yesterday. And the only measurement of this, in turn, is the amount of time that becomes available for the job only human beings can do, the job of communication. By this test, of course, almost no computer today is being used properly. Most of them are being misused, that is, are being used to justify spending even more time on control rather than to relieve human beings from controlling by giving them information. The reason for this is quite clearly the lack of prior communication, that is, of agreement and decision on what information is needed, by whom and for what purposes, and what it means operationally. The reason for the misuse of the computer is, so to speak, the lack of anything comparable to the *Armee Deutsch* of yesterday's much-ridiculed Imperial Austrian Army with its two hundred words of command which even the dumbest recruit could learn in two weeks' time.

The Information Explosion, in other words, is the most impelling reason to go to work on communications. Indeed, the frightening communications gap all around us—between management and workers; between business and government; between faculty and students, and between both of them and university administration; between producers and consumers; and so on—may well reflect in some measure the tremendous increase in information without a commensurate increase in communications.

Can we then say anything constructive about communication? Can we do anything? We can say that communication has to start from the intended recipient of communications rather than from the emitter. In terms of traditional organization we have to start upward. Downward communications cannot work and do not work. They come *after* upward communications have successfully been established. They are reaction rather than action, response rather than initiative.

But we can also say that it is not enough to listen. The upward communication must first be focused on something that both recipient and emitter can perceive, focused on something that is common to both of them. And second, it must be focused on the motivation of the intended recipient. It must, from the beginning, be informed by his values, beliefs, and aspirations.

One example—but only an example: There have been promising results with organizational communication that started out with the demand by the superior that the subordinate think through and present to the superior his own conclusions as to what major contribution to the organization—or to the unit within the organization—the subordinate should be expected to perform and should be held accountable for. What the subordinate then comes up with is rarely what the superior expects. Indeed, the first aim of the exercise is precisely to bring out the divergence in perception between superior and subordinate. But the perception is focused, and focused on something that is real to both parties. To realize that they see the same reality differently is in itself already communication.

Second, in this approach, the intended recipient of communication—in this case the subordinate—is given access to experience that enables him to understand. He is given access to the reality of decision making, the problems of priorities, the choice between what one likes to do and what the situation demands, and above all, the responsibility for a decision. He may not see the situation the same way the superior does—in fact, he rarely will or even should. But he may gain an understanding of the complexity of the superior's situation, and above all of the fact that the complexity is not of the superior's making, but is inherent in the situation itself.

Finally, the communication, even if it consists of a "no" to the subordinate's conclusions, is firmly focused on the aspirations, values, and motivation of the intended recipient. In fact, it starts out with the question, "What would you *want* to do?" It may then end up with the command, "This is what I tell you to do." But at least it forces the superior to realize that he is overriding the desires of the subordinate. It forces him to explain, if not to try to persuade. At least he knows that he has a problem—and so does the subordinate.

A similar approach has worked in another organizational situation in which communication has been traditionally absent: the performance appraisal, and especially the appraisal interview. Performance appraisal is today standard in large organizations (except in Japan, where promotion and pay go by seniority so that performance appraisal would serve little purpose). We know that most people want to know where they stand. One of the most common complaints of employees in organizations is, indeed, that they are not being appraised and are not being told whether they do well or poorly.

The appraisal forms may be filled out. But the appraisal interview in which the appraiser is expected to discuss his performance with a man is almost never conducted. The exceptions are a few organizations in which performance appraisals are considered a communications tool rather than a rating device. This means specifically that the performance appraisal starts out with the question, "What has this man done well?" It then asks, "And what, therefore, should he be able to do well?" And then it asks, "And what would he have to learn or to acquire to be able to get the most from his capacities and achievements?" This, first, focuses on specific achievement. It focuses on things the employee himself is likely to perceive clearly and, in fact, gladly. It also focuses on his own aspirations, values, and desires. Weaknesses are then seen as limitations to what the employee himself can do well and wants to do, rather than as defects. Indeed, the proper conclusion from this approach to appraisal is not the question, "What should the employee do?" but "What should the organization and I, his boss, do?" A proper conclusion is not "What does this communicate to the employee?" It is "What does this communicate to both of us, subordinate *and* superior?"

These are only examples, and rather insignificant ones at that. But perhaps they illustrate conclusions to which our experience with communications—largely an experience of failure—and the work in learning, memory, perception, and motivation point.

The start of communications in organization must be to

get the intended recipient himself to try to communicate. This requires a focus on (a) the impersonal but common task, and (b) on the intended recipient's values, achievements, and aspirations. It also requires the experience of responsibility.

Perception is limited by what can be perceived and geared to what one expects to perceive. Perception, in other words, presupposes experience. Communication within organization, therefore, presupposes that the members of the organization have the foundation of experience to receive and perceive. The artist can convey this experience in symbolical form: he can communicate what his readers or viewers have never experienced. But ordinary managers, administrators, and professors are not likely to be artists. The recipients must, therefore, have actual experience themselves and directly rather than through the vicarious symbols.

Communications in organization demands that the masses, whether they be employees or students, share in the responsibility of decisions to the fullest possible extent. They must understand because they have been through it, rather than accept because it is being explained to them.

I shall never forget the German trade union leader, a faithful Socialist, who was shattered by his first exposure to the deliberations of the Board of Overseers of a large company to which he had been elected as an employee member. That the amount of money available was limited and that, indeed, there was very little money available for all the demands that had to be met, was one surprise. But the pain and complexity of the decisions between various investments, e.g., between modernizing the plant to safeguard workers' jobs and building workers' houses to safeguard their health and family life, was a much bigger and totally unexpected experience. But, as he told me with a half-sheepish, half-rueful grin, the greatest shocker was the realization that all the things he considered important turned out to be irrelevant to the decisions in which he found himself taking an active and responsible part. Yet this man was neither stupid nor dogmatic. He was simply inexperienced—and, therefore, inaccessible to communication.

The traditional defense of paternalism has always been "It's a complex world; it needs the expert, the man who knows best." But paternalism, as our work in perception, learning, and motivation is beginning to bring out, really can work only in a simple world. When people can understand what Papa does because they share his experiences and his perception, then Papa can actually make the decisions for them. In a complex world there is need for a shared experience in the decisions, or there is no common perception, no communications, and, therefore, neither acceptance of the decisions, nor ability to carry them out. The ability to understand presupposes prior communication. It presupposes agreement on meaning.

There will be no communication, in sum, if it is conceived as going from the "I" to the "Thou." Communication only works from one member of "us" to another. Communications in organization—and this may be the true lesson of our communications failure and the true measure of our communications need—are not a *means* of organization. They are a *mode* of organization.

2 | Management's New Role

The major assumptions on which both the theory and the practice of management have been based these past fifty years are rapidly becoming inappropriate. A few of these assumptions are actually no longer valid and, in fact, are obsolete. Others, while still applicable, are fast becoming inadequate; they deal with what is increasingly the secondary, the subordinate, the exceptional, rather than with the primary, the dominant, the ruling function and reality of management. Yet most men of management, practitioners and theoreticians alike, still take these traditional assumptions for granted.

To a considerable extent the obsolescence and inadequacy of these assumed verities of management reflect management's own success. For management has been the success story par excellence of these last fifty years—more so even than science. But to an even greater extent, the traditional assumptions of management scholar and management practitioner are being outmoded by independent—or at least only partially dependent—developments in society, in econ-

Keynote address given at the 15th CIOS International Management Congress, Tokyo, Japan, November 5, 1969.

omy, and in the world view of our age, especially in the developed countries. To a large extent objective reality is changing around the manager—and fast.

Managers everywhere are very conscious of new concepts and new tools of management, of new concepts of organization, for instance, or of the "information revolution." These changes within management are indeed of great importance. But more important yet may be the changes in the basic realities and their impact on the fundamental assumptions underlying management as a discipline and as a practice. The changes in managerial concepts and tools will force managers to change their behavior. The changes in reality demand, however, a change in the manager's role. The changes in concepts and tools mean changes in what a manager *does* and *how* he does it. The change in basic role means a change in what a manager *is*.

The Old Assumptions

Six assumptions may have formed the foundation of the theory and practice of management this last half century. Few practitioners of management have, of course, ever been conscious of them. Even the management scholars have, as a rule, rarely stated them explicitly. But both practitioners and theorists alike have accepted these assumptions, have, indeed, treated them as self-evident axioms and have based their actions, as well as their theories, on them.

These assumptions deal with

- the scope,
- the task,
- the position, and
- the nature of management.

ASSUMPTION ONE: Management is management of business, and business is unique and the exception in society.

This assumption is held subconsciously rather than in full awareness. It is, however, inescapably implied in the view of society which most of us still take for granted whether we are "Right" or "Left," "conservative," "liberal," or "radical," "capitalists" or "communists": the view of European (French and English) seventeenth-century social theory which postulates a society in which there is only one organized power center, the national government, assumed to be sovereign though self-limited, with the rest of society essentially composed of the social molecules of individual families. Business, if seen at all in this view, is seen as the one exception, the one organized institution. Management, therefore, is seen as confined to the special, the atypical, the isolated institution of the economic sphere, i.e., to business enterprise. The nature as well as the characteristics of management are, in the traditional view, thus very largely grounded in the nature and the characteristics of business activity. One is "for management" if one is "for business," and vice versa. And somehow, in this view, economic activity is quite different from all other human concerns—to the point where it has become fashionable to speak of "the economic concern" as opposed to "human concerns."

ASSUMPTION TWO: "Social responsibilities" of management, that is, concerns that cannot be encompassed within an economic calculus, are restraints and limitations imposed on management rather than management objectives and tasks. They are to be discharged largely without the enterprise and outside of management's normal working day. At the same time and because business is assumed to be the one exception, only business has social responsibilities; indeed, the common phrase is "the social responsibilities of business."

University, hospital, or government agencies are clearly not assumed, in the traditional view, to have any social responsibilities.

This view derives directly from the belief that business is the one, the exceptional institution. University and hospital are not assumed to have any social responsibility primarily because they are not within the purview of the traditional vision—they are simply not seen at all as "organizations." Moreover, the traditional view of a social responsibility peculiar, and confined, to business derives from the assumptions that economic activity differs drastically from other human activities (if, indeed, it is even seen as a "normal" human activity), and that "profit" is something extraneous to the economic process and imposed on it by the "capitalist" rather than an intrinsic necessity of any economic activity whatever.

ASSUMPTION THREE: The primary, perhaps the only, task of management is to mobilize the energies of the business organization for the accomplishment of known and defined tasks. The tests are efficiency in doing what is already being done, and adaptation to changes outside. Entrepreneurship and innovation—other than systematic research—lie outside the management scope.

To a large extent this assumption was a necessity during the last half century. The new fact then was, after all, not entrepreneurship and innovation with which the developed countries had been living for several hundred years. The new fact of the world of 1900, when concern with management first arose, was the large and complex organization for production and distribution with which the traditional managerial systems, whether of workshop or of local store, could not cope. The invention of the steam locomotive was not what triggered concern with management. Rather it was the emergence, some fifty years later, of the large railroad company which could handle steam locomotives without much trouble but was baffled by the problem of co-ordination between people, of communication between them, and of their authorities and responsibilities.

But the focus on the *managerial* side of management—to the almost total neglect of entrepreneurship as a function of management—also reflects the reality of the economy in the half century since World War II. It was a period of high technological and entrepreneurial continuity, a period that required adaptation rather than innovation, and ability to do better rather than courage to do differently.*

The long and hard resistance against management on the part of the German *Unternehmer* or French *patron* reflects in large measure a linguistic misunderstanding. There is no German or French word that adequately renders *management*, just as there is no English word for *entrepreneur* (which has remained a foreigner after almost two hundred years of sojourn in the English-speaking world). In part this resistance arises out of peculiarities of economic structure, e.g., the role of the commercial banks in Germany which makes the industrialist concerned for his autonomy stress the "charisma" of the *Unternehmer* against the impersonal professionalism of the "manager." In part also "management" is classless and derives its authority from its objective function rather than, as does German *Unternehmer* or French *patron*, from ownership or social class. But surely one of the main reasons for the resistance against "management" —both as a term and as a concept—on the continent of Europe has been the—largely subconscious—emphasis on

* As documented in some detail in my book, *The Age of Discontinuity* (New York, Harper & Row, 1969).

the managerial internal task as against the external, entre-preneurial, innovating function.

ASSUMPTION FOUR: It is the manual worker—skilled or un-skilled—who is management's concern as resource, as a cost center, and as a social and individual problem.

To have made the manual worker productive is, indeed, the greatest achievement of management to date. Frederick Winslow Taylor's "Scientific Management" is often at-tacked these days (though mostly by people who have not read Taylor). But it was his insistence on studying work that underlies the affluence of today's developed coun-tries; it raised the productivity of manual work to the point where yesterday's "laborer"—a proletarian con-demned to an income at the margin of subsistence by "the iron law of wages" and to complete uncertainty of em-ployment from day to day—has become the "semiskilled worker" of today's mass production industries with a middle-class standard of living and guaranteed job or income security. And Taylor thereby found the way out of the apparently hopeless impasse of nineteenth-century "class war" between the "capitalist exploitation" of the laboring man and the "proletarian dictatorship."

As late as World War II, the central concern was still the productivity and management of manual work; the central achievement of both the British and the American war economies was the mobilization, training, and manag-ing of production workers in large numbers. Even in the postwar period one major task—in all developed countries other than Great Britain—was the rapid conversion of immigrants from the farm into productive manual workers in industry. On this accomplishment—made possible only because of the "Scientific Management" which Taylor pio-neered seventy years ago—the economic growth and performance of Japan, of Western Europe, and even of the United States largely rest.

ASSUMPTION FIVE: Management is a "science" or at least a "discipline," that is, it is as independent of cultural values and individual beliefs as are the elementary operations of arithmetic, the laws of physics, or the stress tables of the engineer. But all management is being practiced within one distinct national environment and imbedded in one national culture, circumscribed by one legal code and part of one national economy.

These two propositions were as obvious to Taylor in the United States as they were to Fayol in France. Of all the early management authorities only Rathenau in Germany seemed to have doubted that management was an "objective," i.e., culture-free, discipline—and no one listened to him. The Human Relations school attacked Taylor as "unscientific"; they did not attack Taylor's premise that there was an objective "science" of management. On the contrary, the Human Relations school proclaimed its findings to be "true" scientific psychology and grounded in the "Nature of Man." It refused even to take into account the findings of its own colleagues in the social sciences, the cultural anthropologists. Insofar as cultural factors were considered at all in the traditional assumptions of management, they were "obstacles." It is still almost axiomatic in management that social and economic development requires the abandonment of "nonscientific," i.e., traditional cultural beliefs, values, and habits. And the Russians, e.g., in their approach to Chinese development under Stalin, differ no whit from the Americans or the Germans in respect to this assumption. That, however, the assumption is little but Western narrowness and cultural egocentricity, one look at the development of Japan would have shown.

At the same time, management theory and practice saw in the national state and its economy the "natural" habitat of business enterprise—as did (and still does), of course, all our political, legal, and economic theory.

ASSUMPTION SIX: Management is the result of economic development.

This had, of course, been the historical experience in the West (though not in Japan where the great organizers such as Mitsui, Iwasaki, Shibusawa, came first, and where economic development, clearly, was the result of management). But even in the West the traditional explanation of the emergence of management was largely myth. As the textbooks had it (and still largely have it), management came into being when the small business outgrew the owner who had done everything himself. In reality, management evolved in enterprises that started big and could never have been anything but big—the railways in particular, but also the postal service, the steamship companies, the steel mills, and the department stores. To industries that could start small, management came very late; some of those, e.g., the textile mill or the bank, are still often run on the pattern of the "one boss" who does everything and who, at best, has "helpers." But even where this was seen—and Fayol as well as Rathenau apparently realized that management was a function rather than a stage—management was seen as a result rather than as a cause, and as a response to needs rather than as a creator of opportunity.

*   *   *

I fully realize that I have oversimplified—grossly so. But I do not believe that I have misrepresented our traditional assumptions. Nor do I believe that I am mistaken that these assumptions, in one form or another, still underlie both the theory and the practice of management, especially in the industrially developed nations.

—And the New Realities

Today, however, we need quite different assumptions. They, too, of course, oversimplify—and grossly, too. But they are far closer to today's realities than the assumptions on which theory and practice of management have been basing themselves these past fifty years.

Here is a first attempt to formulate assumptions that correspond to the management realities of our time.

ASSUMPTION ONE: Every major task of developed society is being carried out in and through an organized and managed institution. Business enterprise was only the first of those and, therefore, became the prototype by historical accident. But while it has a specific job—the production and distribution of economic goods and services—it is neither the exception nor unique. Large-scale organization is the rule rather than the exception. Our society is one of pluralist organizations rather than a diffusion of family units. And management, rather than the isolated peculiarity of one unique exception, the business enterprise, is generic and the central social function in our society.*

A recent amusing book† points out that management is a form of government and applies to it Machiavelli's classic insights. But this is really not a very new idea, far

---

* On this New Pluralism, see *The Age of Discontinuity*, especially Part Three: "A Society of Organizations."

† Anthony Jay: *Management and Machiavelli* (New York: Holt Rinehart and Winston, 1968).

from it. It underlay a widely read book of 1941—James Burnham's *Managerial Revolution* (though Mr. Burnham applied Marx rather than Machiavelli to management). It was treated in considerable detail in three of my books— *The Future of Industrial Man* (1942), *Concept of the Corporation* (1946), and *The New Society* (1950). Mr. Justice Brandeis knew this well before World War I when he coined the term "Scientific Management" for Frederick Taylor's investigations into manual work. That a business organization is a form of government was also perfectly obvious to the entire tradition of American institutional economics from John R. Commons on, that is, from around the turn of the century. And across the Atlantic, Walter Rathenau saw the same thing clearly well before 1920.

But what is new is that nonbusiness institutions flock in increasing numbers to business management to learn from it how to manage themselves. The hospital, the armed services, the Catholic diocese, the civil service—all want to go to school for business management. And where Britain's first postwar Labor government nationalized the Bank of England to prevent its being run like a business, the next Labor government hired in 1968 a leading American firm of management consultants (McKinsey & Company) to reorganize the Bank of England to make sure it would be managed as a business.

This does not mean that business management can be transferred to other, nonbusiness institutions. On the contrary, the first thing these institutions have to learn from business management is that management begins with the setting of objectives and that, therefore, noneconomic institutions, such as a university or a hospital, will also need very different management from that of a business. But these institutions are right in seeing in business management the prototype. What we have done in respect to the management of a business we increasingly will have to do for the other institutions, including the government agencies. Business, far from being exceptional, is, in other words, simply the first of the species and the one we have studied the most intensively. And management is generic rather than the exception.

Indeed, what has always appeared as the most exceptional characteristic of business management, namely, the measurement of results in economic terms, that is, in terms of profitability, now emerges as the exemplar of what all institutions need: an objective outside measurement of the allocation of resources to results and of the rationality of managerial decisions. Noneconomic institutions need a yardstick that does for them what profitability does for the business—this underlies the attempt of Robert McNamara, while Secretary of Defense of the United States, to introduce "cost-effectiveness" into the government and to make planned, purposeful, and continuous measurement of programs by their results, as compared to their promises and expectations, the foundation for budget and policy decisions. "Profitability," in other words, rather than being the "exception" and distinct from "human" or "social" needs, emerges, in the pluralist society of organizations, as the prototype of the measurement needed by every institution to be managed and manageable.*

ASSUMPTION TWO: Because our society is rapidly becoming a society of organizations, all institutions, including business, will have to hold themselves accountable for the "quality of life" and will have to make fulfillment of basic social values, beliefs, and purposes a major objective of their continuing normal activities rather than a "social responsibility" that

---

* This, of course, also underlies the rapid return to profit and profitability as yardsticks and determinants of allocation decisions in the developed communist countries, that is, Russia and the European satellites.

restrains or that lies outside of their normal main functions. They will have to learn to make the "quality of life" into an opportunity for their own main tasks. In the business enterprise, this means that the attainment of the "quality of life" increasingly will have to be considered a business opportunity and will have to be converted by management into profitable business.

This will apply increasingly to fulfillment of the individual. It is the organization which is today our most visible social environment. The family is "private" rather than "community"—not that this makes it any less important. The "community" is increasingly in the organization, and especially in the one in which the individual finds his livelihood and through which he gains access to function, achievement, and social status. (On this see my *The New Society* [New York, 1950].) It will increasingly be the job of management to make the individual's values and aspirations redound to organizational energy and performance. It will simply not be good enough to be satisfied—as Industrial Relations and even Human Relations traditionally have been—with "satisfaction," that is, with the absence of discontent. Perhaps one way to dramatize this is to say that we will, within another ten years, become far less concerned with "management development" as a means of adapting the individual to the demands of the organization and far more with "organization development" to adapt the organization to the needs, aspirations, and potential of the individual.

ASSUMPTION THREE: Entrepreneurial innovation will be as important to management as the managerial function, both in the developed and in the developing countries. Indeed, entrepreneurial innovation may be more important in the years to come. Unlike the nineteenth century, however, entrepreneurial innovation will increasingly have to be carried out in and by existing institutions such as existing businesses. It will, therefore, no longer be possible to consider it as lying outside of management or even as peripheral to management. Entrepreneurial innovation will have to become the very heart and core of management.

There is every reason to believe* that the closing decades of the twentieth century will see changes as rapid as those that characterized the fifty years between 1860 and 1914, when a new major invention ushering in almost immediately a new major industry with new big businesses appeared on the scene every two to three years. But unlike the last century, these innovations of our century will be as much social innovations as they will be technical; a metropolis, for instance, is clearly as much of a challenge to the innovator today as the new science of electricity was to the inventor of 1870. And unlike the last century, innovation in this century will be based increasingly on knowledge of any kind rather than on science alone.

At the same time, innovation will increasingly have to be channeled in and through existing businesses, if only because the tax laws in every developed country make the existing business the center of capital accumulation. And innovation is capital-intensive, especially in the two crucial phases, the development phase and the market introduction of new products, processes, or services. We will, therefore, increasingly have to learn to make existing organizations capable of rapid and continuing innovation. How far we are still from this is shown by the fact that management still worries about "resistance to change." What existing organizations will have to learn is to reach out for change as an opportunity and to resist continuity.

* For documentation, see *The Age of Discontinuity*.

ASSUMPTION FOUR: A primary task of management in the developed countries in the decades ahead will increasingly be to make knowledge productive. The manual worker is "yesterday"—and all we can fight on that front is a rear-guard action. The basic capital resource, the fundamental investment, but also the cost center of a developed economy, is the knowledge worker who puts to work what he has learned in systematic education, that is, concepts, ideas, and theories, rather than the man who puts to work manual skill or muscle.

Taylor put knowledge to work to make the manual worker productive. But Taylor himself never asked the question, "What constitutes 'productivity' in respect to the industrial engineer who applies 'Scientific Management'?" As a result of Taylor's work, we can answer what productivity is in respect to the manual worker. But we still cannot answer what productivity is in respect to the industrial engineer, or to any other knowledge worker. Surely the measurements which give us productivity for the manual worker, such as the number of pieces turned out per hour or per dollar of wage, are quite irrelevant if applied to the knowledge worker. There are few things as useless and unproductive as the engineering department which with great dispatch, industry, and elegance turns out the drawings for an unsalable product. Productivity in respect to the knowledge worker is, in other words, primarily quality. We cannot even define it yet.

One thing is clear: to make knowledge productive will bring about changes in job structure, careers, and organizations as drastic as those which resulted in the factory from the application of Scientific Management to manual work. The entrance job, above all, will have to be changed drastically to enable the knowledge worker to become productive. For it is abundantly clear that knowledge cannot be productive unless the knowledge worker finds out who he himself is, what kind of work he is fitted for, and how he best works. In other words, there can be no divorce of "planning" from "doing" in knowledge work. On the contrary, the knowledge worker must be able to "plan" himself. And this the present entrance jobs, by and large, do not make possible. They are based on the assumption—valid for manual work but quite inappropriate to knowledge work—that an outsider can objectively determine the "one best way" for any kind of work. For knowledge work, this is simply not true. There may be "one best way," but it is heavily conditioned by the individual and not entirely determined by physical, or even by mental, characteristics of the job. It is temperamental as well.

ASSUMPTION FIVE: There are management tools and techniques. There are management concepts and principles. There is a common language of management. And there may even be a universal "discipline" of management. Certainly there is a worldwide generic function which we call management and which serves the same purpose in any and all developed societies. But management is also a culture and a system of values and beliefs. It is also the means through which a given society makes productive its own values and beliefs. Indeed, management may well be considered the bridge between a "civilization" that is rapidly becoming worldwide, and a "culture" which expresses divergent tradition, values, beliefs, and heritages. Management must, indeed, become the instrument through which cultural diversity can be made to serve the common purposes of mankind. At the same time, management increasingly is not being practiced within the confines of one national culture, law, or sovereignty but "multinationally." Indeed, management increasingly is becoming an institution—so far, the only one—of a genuine world economy.

Management we now know has to make productive values, aspirations, and traditions of individuals, community, and society for a common productive purpose. If management, in other words, does not succeed in putting to work the specific cultural heritage of a country and of a people, social and economic development cannot take place. This is, of course, the great lesson of Japan—and the fact that Japan succeeded, a century ago, in putting to work her own traditions of community and human values for the new ends of a modern industrialized state explains why Japan succeeded while every other non-Western country has so far failed. Management, in other words, will have to be considered both a science and a humanity, both a statement of findings that can be objectively tested and validated and a system of belief and experience.

At the same time, management—and here we speak of business management alone, so far—is rapidly emerging as the one and only institution that is common and transcends the boundaries of the national state. The multinational corporation does not really exist so far. What we have, by and large, are still businesses that are based on one country with one culture and, heavily, one nationality, especially in top management. But it is also becoming clear that this is a transition phenomenon and that continuing development of the world economy both requires and leads to genuinely multinational companies in which not only production and sales are multinational, but ownership and management as well—all the way from the top down.

Within the individual country, especially the developed country, business is rapidly losing its exceptional status as we recognize that it is the prototype of the typical, indeed, the universal, social form, the organized institution requiring management. Beyond the national boundary, however, business is rapidly acquiring the same exceptional status it no longer has within the individual developed country. Beyond the national boundary, business is rapidly becoming the unique, the exceptional, the one institution which expresses the reality of a world economy and of a worldwide knowledge society.

ASSUMPTION SIX: Management creates economic and social development. Economic and social development is the *result* of management.

It can be said without too much oversimplification that there are no "underdeveloped countries." There are only "undermanaged" ones. Japan a hundred years ago was an underdeveloped country by every material measurement. But it very quickly produced management of great competence, indeed, of excellence. Within twenty-five years Meiji Japan had become a developed country, and, indeed, in some aspects, such as literacy, the most highly developed of all countries. We realize today that it is Meiji Japan, rather than eighteenth-century England—or even nineteenth-century Germany—which has to be the model of development for the underdeveloped world. This means that management is the prime mover and that development is a consequence.

All our experience in economic development proves this. Wherever we have only contributed the economic "factors of production," especially capital, we have not achieved development. In the few cases where we have been able to generate management energies (e.g., in the Cauca Valley in Colombia*) we have generated rapid

---

* For a description, see my *Age of Discontinuity*, pp. 129–30.

development. Development, in other words, is a matter of human energies rather than of economic wealth. And the generation and direction of human energies is the task of management.

                    ❋   ❋   ❋

Admittedly, these new assumptions oversimplify; they are meant to. But I submit that they are better guides to effective management in the developed countries today, let alone tomorrow, than the assumptions on which we have based our theories as well as our practice these last fifty years. We are not going to abandon the old tasks. We still, obviously, have to manage the going enterprise and have to create internal order and organization. We still have to manage the manual worker and make him productive. And no one who knows the reality of management is likely to assert that we know everything in these and similar areas that we need to know; far from it. But the big jobs waiting for management today, the big tasks requiring both new theory and new practice, arise out of the new realities and demand different assumptions and different approaches.

More important even than the new tasks, however, may be management's new role. Management is fast becoming the central resource of the developed countries and the basic need of the developing ones. From being the specific concern of one, the economic institutions of society, management and managers are becoming the generic, the distinctive, the constitutive organ of developed society. What management is and what managers do will, therefore—and properly—become increasingly a matter of public concern rather than a matter for the "experts." Management will increasingly be concerned as much with the expression of basic beliefs and values as with the accomplishment of measurable results. It will increasingly stand for the quality of life of a society as much as for its standard of living.

There are many new tools of management the use of which we will have to learn, and many new techniques. There are, as this paper points out, a great many new and difficult tasks. But the most important change ahead for management is that increasingly the aspirations, the values, indeed, the very survival of society in the developed countries will come to depend on the performance, the competence, the earnestness and the values of their managers. The task of the next generation is to make productive for individual, community, and society the new organized institutions of our New Pluralism. And that is, above all, the task of management.

3 | Work and Tools

Man, alone of all animals, is capable of purposeful, nonorganic evolution; he makes tools. This observation by Alfred Russell Wallace, codiscoverer with Darwin of the theory of evolution, may seem obvious if not trite. But it is a profound insight. And though made some seventy or eighty years ago, its implications have yet to be thought through by biologists and technologists.

One such implication is that from a biologist's (or a historian's) point of view, the technologist's identification of tool with material artifact is quite arbitrary. Language, too, is a tool, and so are all abstract concepts. This does not mean that the technologist's definition should be discarded. All human disciplines rest after all on similarly arbitrary distinctions. But it does mean that technologists ought to be conscious of the artificiality of their definition and careful lest it become a barrier rather than a help to knowledge and understanding.

This is particularly relevant for the history of technology, I believe. According to the technologist's definition of "tool," the abacus and the geometer's compass are normally con-

First published in *Technology and Culture*, Winter, 1959.

sidered technology, but the multiplication table or table of logarithms is not. Yet this arbitrary division makes all but impossible the understanding of so important a subject as the development of the technology of mathematics. Similarly the technologist's elimination of the fine arts from his field of vision blinds the historian of technology to an understanding of the relationship between scientific knowledge and technology. (See, for instance, Volumes 3 and 4 of Singer's monumental *History of Technology*.) For scientific thought and knowledge were married to the fine arts, at least in the West, long before they even got on speaking terms with the mechanical crafts: in the mathematical number theories of the designers of the Gothic cathedral,* in the geometric optics of Renaissance painting, or in the acoustics of the great Baroque organs. And Lynn T. White, Jr., has shown in several recent articles that to understand the history and development of the mechanical devices of the Middle Ages we must understand something so nonmechanical and nonmaterial as the new concept of the dignity and sanctity of labor which St. Benedict first introduced.

Even within the technologist's definition of technology as dealing with mechanical artifacts alone, Wallace's insight has major relevance. The subject matter of technology, according to the Preface to *History of Technology*, is "how things are done or made"; and most students of technology, to my knowledge, agree with this. But the Wallace insight leads to a different definition: the subject matter of technol-

---

* S. B. Hamilton only expresses the prevailing view of technologists when he says (in Singer's *History of Technology*, IV, 469) in respect to the architects of the Gothic cathedral and their patrons that there is "nothing to suggest that either party was driven or pursued by any theory as to what would be beautiful." Yet we have overwhelming and easily accessible evidence to the contrary; both architect and patron were not just "driven," they were actually obsessed by rigorously mathematical theories of structure and beauty. See, for instance, Sedlmayer, *Die Entstehung der Kathedrale* (Zurich, 1950); Von Simson, *The Gothic Cathedral* (New York, 1956); and especially the direct testimony of one of the greatest of the cathedral designers, Abbot Suger of St. Denis, in *Abbot Suger and the Abbey Church of St. Denis*, ed. Erwin Panofsky (Princeton, 1946).

ogy would be "how man does or makes." As to the meaning
and end of technology, the same source, again presenting the
general view, defines them as "mastery of his (man's) natural environment." Oh no, the Wallace insight would say
(and in rather shocked tones): the purpose is to overcome
man's own natural, i.e., animal, limitations. Technology enables man, a land-bound biped, without gills, fins, or wings,
to be at home in the water or in the air. It enables an animal
with very poor body insulation, that is, a subtropical animal,
to live in all climate zones. It enables one of the weakest and
slowest of the primates to add to his own strength that of
elephant or ox, and to his own speed that of the horse. It
enables him to push his life span from his "natural" twenty
years or so to threescore years and ten; it even enables him
to forget that natural death is death from predators, disease,
starvation, or accident, and to call death from natural causes
that which has never been observed in wild animals: death
from organic decay in old age.*

These developments of man have, of course, had impact
on his natural environment—though I suspect that until
recent days the impact has been very slight indeed. But this
impact on nature outside of man is incidental. What really
matters is that all these developments alter man's biological
capacity—and not through the random genetic mutation of
biological evolution but through the purposeful nonorganic
development we call technology.

What I have called here the "Wallace insight," that is, the
approach from human biology, thus leads to the conclusion
that technology is not about things: tools, processes, and
products. It is about work: the specifically human activity
by means of which man pushes back the limitations of the
iron biological law which condemns all other animals to
devote all their time and energy to keeping themselves alive
for the next day, if not for the next hour. The same conclu-

* See on this P. B. Medawar, the British biologist, in "Old Age and Natural Death" in his *The Uniqueness of the Individual* (New York, 1957).

sion would be reached, by the way, from any approach, for instance, from that of the anthropologist's "culture," that does not mistake technology for a phenomenon of the physical universe. We might define technology as human action on physical objects or as a set of physical objects characterized by serving human purposes. Either way the realm and subject matter of the study of technology would be human work.

For the historian of technology this line of thought might be more than a quibble over definitions. For it leads to the conclusion that the study of the development and history of technology, even in its very narrowest definition as the study of one particular mechanical artifact (either tool or product) or a particular process, would be productive only within an understanding of work and in the context of the history and development of work.

Not only must the available tools and techniques strongly influence what work can and will be done, but how it will be done. Work, its structure, organization, and concepts must in turn powerfully affect tools and techniques and their development. The influence, one would deduce, should be so great as to make it difficult to understand the development of the tool or of the technique unless its relationship to work was known and understood. Whatever evidence we have strongly supports this deduction.

Systematic attempts to study and to improve work only began some seventy-five years ago with Frederick W. Taylor. Until then work had always been taken for granted by everyone—as it is still, apparently, taken for granted by most students of technology. "Scientific management," as Taylor's efforts were called misleadingly ("scientific work study" would have been a better term and would have avoided a great deal of confusion), was not concerned with technology. Indeed, it took tools and techniques largely as given and tried to enable the individual worker to manipulate them more economically, more systematically, and more effectively. And yet this approach resulted almost immediately in major changes and development in tools, processes, and products. The assembly line with its conveyors was an important tool change. An even greater change was the change in process that underlay the switch from building to assembling a product. Today we are beginning to see yet another powerful consequence of Taylor's work on individual operations: the change from organizing production around the doing of things to things to organizing production around the flow of things and information, the change we call "automation."

A similar, direct impact on tools and techniques is likely to result from another and even more recent approach to the study and improvement of work: the approach called variously "human engineering," "industrial psychology," or "industrial physiology." Scientific Management and its descendants study work as operation; human engineering and its allied disciplines are concerned with the relationship between technology and human anatomy, human perception, human nervous system, and human emotion. Fatigue studies were the earliest and most widely known examples; studies of sensory perception and reaction, for instance, of airplane pilots, are among the presently most active areas of investigation, as are studies of learning. We have barely scratched the surface here; yet we know already that these studies are leading us to major changes in the theory and design of instruments of measurement and control, and into the redesign of traditional skills. traditional tools, and traditional processes.

But of course we worked on work, if only through trial and error, long before we systematized the job. The best example of Scientific Management is after all not to be found in our century: it is the alphabet. The assembly line as a concept of work was understood by those unknown geniuses who, at the very beginning of historical time, replaced the aristocratic artist of warfare (portrayed in his last moments of glory by Homer) by the army soldier with his uniform equipment, his few repetitive operations, and his regimented drill. The best example of human engineering is still the long handle that changed the sickle into the scythe, thus belatedly adjusting reaping to the evolutionary change that had much earlier changed man from crouching quadruped into upright biped. Every one of these developments in work had immediate and powerful impact on tools, process, and product, that is, on the artifacts of technology.

The aspect of work that has probably had the greatest impact on technology is the one we know the least about: the organization of work.

Work, as far back as we have any record of man, has always been both individual and social. The most thoroughly collectivist society history knows, that of Inca Peru, did not succeed in completely collectivizing work; technology—in particular, the making of tools, pottery, textiles, and cult objects—remained the work of individuals. It was personally specialized rather than biologically or socially specialized, as is work in a beehive or in an ant heap. The most thoroughgoing individualist society, the perfect market model of classical economics, presupposed a tremendous amount of collective organization in respect to law, money and credit, transportation, and so on. But precisely because individual effort and collective effort must always be calibrated with one another, the organization of work is not determined. To a very considerable extent there are genuine alternatives here, genuine choices. The organization of work, in other words, is in itself one of the major means of that purposeful and nonorganic evolution which is specifically human; it is in itself an important tool of man.

Only within the very last decades have we begun to look at the organization of work.* But we have already learned that the task, the tools, and the social organization of work are not totally independent but mutually influence and affect one another. We know, for instance, that the almost preindustrial technology of the New York women's dress industry is the result not of technological, economic, or market conditions but of the social organization of work which is traditional in that industry. The opposite has been proven, too: When we introduce certain tools into locomotive shops, for instance, the traditional organization of work, the organization of the crafts, becomes untenable; and the very skills that made men productive under the old technology now become a major obstacle to their being able to produce at all. A good case can be made for the hypothesis that modern farm implements have made the Russian collective farm socially obsolete as an organization of work, have made it yesterday's socialist solution of farm organization rather than today's, let alone tomorrow's.

This interrelationship between organization of work, tasks, and tools must always have existed. One might even speculate that the explanation for the mysterious time gap between the early introduction of the potter's wheel and the so very late introduction of the spinning wheel lies in the

---

* Among the studies ought to be mentioned the work of the late Elton Mayo, first in Australia and then at Harvard, especially his two slim books: *The Human Problems of an Industrial Civilization* (Boston, 1933) and *The Social Problems of an Industrial Civilization* (Boston, 1945); the studies of the French sociologist Georges Friedmann, especially his *Industrial Society* (Glencoe, Illinois, 1955); the work carried on at Yale by Charles Walker and his group, especially the book by him and Robert H. Guest: *The Man on the Assembly Line* (Cambridge, Massachusetts, 1952). I understand that studies of the organization of work are also being carried out at the Polish Academy of Science, but I have not been able to obtain any of the results.

social organization of spinning work as a group task performed, as the Homeric epics describe it, by the mistress
working with her daughters and maids. The spinning wheel
with its demand for individual concentration on the machinery and its speed is hardly conducive to free social intercourse; even on a narrowly economic basis, the governmental, disciplinary, and educational yields of the spinning
bee may well have appeared more valuable than faster and
cleaner yarn.

If we know far too little about work and its organization
scientifically, we know nothing about it historically. It is not
lack of records that explains this, at least not for historic
times. Great writers—Hesiod, Aristophanes, Vergil, for instance—have left detailed descriptions. For the early empires and then again for the last seven centuries, beginning
with the High Middle Ages, we have an abundance of
pictorial material: pottery and relief paintings, woodcuts,
etchings, prints. What is lacking is attention and objective
study.

The political historian or the art historian, still dominated by
the prejudices of Hellenism, usually dismisses work as beneath his notice; the historian of technology is "thing-
focused." As a result, we not only still repeat as fact traditions regarding the organization of work in the past which
both our available sources and our knowledge of the organization of work would stamp as old wives' tales. We also
deny ourselves a fuller understanding of the already existing
and already collected information regarding the history and
use of tools.

One example of this is the lack of attention given to
materials-moving and materials-handling equipment. We
know that moving things—rather than fabricating things—is
the central effort in production. But we have paid little

attention to the development of materials-moving and materials-handling equipment.

The Gothic cathedral is another example. H. G. Thomson in *History of Technology* (II, 384) states flatly, for instance, "there was no exact medieval equivalent of the specialized architect" in the Middle Ages; there was only "a master mason." But we have overwhelming evidence to the contrary (summarized, for instance, in Simson*); the specialized, scientifically trained architect actually dominated. He was sharply distinguished from the master mason by training and social position. Far from being anonymous, as we still commonly assert, he was a famous man, sometimes with an international practice ranging from Scotland to Poland to Sicily. Indeed, he took great pains to make sure that he would be known and remembered, not only in written records but above all by having himself portrayed in the churches he designed in his full regalia as a scientific geometer and designer—something even the best-known of today's architects would hesitate to do. Similarly we still repeat early German Romanticism in the belief that the Gothic cathedral was the work of individual craftsmen. But the structural fabric of the cathedral was based on strict uniformity of parts. The men worked to molds which were collectively held and administered as the property of the guild. Only roofing, ornaments, doors, statuary, windows, and so on, were individual artists' work. Considering both the extreme scarcity of skilled people and the heavy dependence on local, unskilled labor from the countryside to which all our sources attest, there must also have been a sharp division between the skilled men who made parts and the unskilled who assembled them under the direction of a foreman or a gang boss. There must thus have been a fairly advanced materials-handling technology which is, indeed,

---

* Von Simson, *The Gothic Cathedral* (New York, 1956) pp. 30 ff.

depicted in our sources but neglected by the historians of technology with their uncritical Romanticist bias. And while the molds to which the craftsman worked are generally mentioned, no one, to my knowledge, has yet investigated so remarkable a tool, and one that so completely contradicts all we otherwise believe we know about medieval work and technology.

I do not mean to suggest that we drop the historical study of tools, processes, and products. We quite obviously need to know much more. I am saying first that the history of work is in itself a big, rewarding, and challenging area which students of technology should be particularly well equipped to tackle. I am saying also that we need work on work if the history of technology is truly to be history and not just the engineer's antiquarianism.

One final question must be asked: Without study and understanding of work, how can we hope to arrive at an understanding of technology?

Singer's great *History of Technology* abandons the attempt to give a comprehensive treatment of its subject with 1850; at that time, the editors tell us, technology became so complex as to defy description, let alone understanding. But it is precisely then that technology began to be a central force and to have major impact both on man's culture and on man's natural environment. To say that we cannot encompass modern technology is very much like saying that medicine stops when the embryo issues from the womb. We need a theory that enables us to organize the variety and complexity of modern tools around some basic, unifying concept.

To a layman who is neither professional historian nor professional technologist, it would, moreover, appear that even the old technology, the technology before the great explosion of the last hundred years, makes no real sense and cannot be understood, can hardly even be described, without such basic concepts. Every writer on technology acknowledges the extraordinary number, variety, and complexity of factors that play a part in technology and are in turn influenced by it: economy and legal system, political institutions and social values, philosophical abstractions, religious beliefs, and scientific knowledge. No one can know all these, let alone handle them all in their constantly shifting relationship. Yet all of them are part of technology in one way or another, at one time or another.

The typical reaction to such a situation has of course always been to proclaim one of these factors as *the* determinant—the economy, for instance, or the religious beliefs. We know that this can only lead to complete failure to understand. These factors profoundly influence but do not determine each other; at most they may set limits to each other or create a range of opportunities for each other. Nor can we understand technology in terms of the anthropologist's concept of culture as a stable, complete, and finite balance of these factors. Such a culture may exist among small, primitive, decaying tribes, living in isolation. But this is precisely the reason why they are small, primitive, and decaying. Any viable culture is characterized by capacity for internal self-generated change in the energy-level and direction of any one of these factors and in their interrelationships.

Technology, in other words, must be considered as a system,* that is, a collection of interrelated and intercommunicating units and activities.

We know that we can study and understand such a system only if we have a unifying focus where the interaction of *all* the forces and factors within the system registers some

---

* The word is here used as in Kenneth Boulding's "General Systems Theory—The Skeleton of Science," *Management Science,* II, No. 3 (April 1956), 197, and in the publications of the Society for General Systems Research.

discernible effect, and where in turn the complexities of the system can be resolved in one theoretical model. Tools, processes, and products are clearly incapable of providing such focus for the understanding of the complex system we call technology. It is just possible, however, that work might provide the focus, might provide the integration of all these interdependent, yet autonomous variables, might provide one unifying concept which will enable us to understand technology both in itself and in its role, its impact on and relationships with values and institutions, knowledge and beliefs, individual and society.

Such understanding would be of vital importance today. The great, perhaps the central, event of our times is the disappearance of all non-Western societies and cultures under the inundation of Western technology. Yet we have no way of analyzing this process, of predicting what it will do to man, his institutions and values, let alone of controlling it, that is, of specifying with any degree of assurance what needs to be done to make this momentous change productive or at least bearable. We desperately need a real understanding, and a real theory, a real model of technology.

History has never been satisfied to be a mere inventory of what is dead and gone—that, indeed, is antiquarianism. True history always aims at helping us understand ourselves, at helping us make what shall be. Just as we look to the historian of government for a better understanding of government, and to the historian of art for a better understanding of art, so we are entitled to look to the historian of technology for a better understanding of technology. But how can he give us such an understanding unless he himself has some concept of technology and not merely a collection of individual tools and artifacts? And can he develop such a concept unless work rather than things becomes the focus of his study of technology and of its history?

4 | Technological Trends
in the Twentieth Century

Technological activity during the twentieth century has changed in its structure, methods, and scope. It is this qualitative change which explains even more than the tremendous rise in the volume of work the emergence of technology in the twentieth century as central in war and peace, and its ability within a few short decades to remake man's way of life all over the globe.

This over-all change in the nature of technological work during this century has three separate though closely related aspects: (1) structural changes—the professionalization, specialization, and institutionalization of technological work; (2) changes in methods—the new relationship between technology and science; the emergence of systematic research; and the new concept of innovation; and (3) the "systems approach." Each of these is an aspect of the same fundamental trend. Technology has become what it never was before: an organized and systematic discipline.

First published in vol. 2 of *Technology in Western Civilization,* ed. Melvin Kranzberg and Carroll W. Pursell, Jr. (New York: Oxford University Press, 1967).

The Structure of Technological Work

Throughout the nineteenth century technological activity, despite tremendous success, was still in its structure almost entirely what it had been through the ages: a craft. It was practiced by individuals here, there, and yonder, usually working alone and without much formal education. By the middle of the twentieth century technological activity has become thoroughly professional, based, as a rule, on specific university training. It has become largely specialized, and is to a very substantial extent being carried out in a special institution—the research laboratory, particularly the industrial research laboratory—devoted exclusively to technological innovation.

Each of these changes deserves a short discussion. To begin with, few of the major figures in nineteenth-century technology received much formal education. The typical inventor was a mechanic who began his apprenticeship at age fourteen or earlier. The few who had gone to college had not, as a rule, been trained in technology or science but were liberal arts students, trained primarily in classics. Eli Whitney (1765–1825) and Samuel Morse (1791–1873), both Yale graduates, are good examples. There were, of course, exceptions such as the Prussian engineering officer Werner von Siemens (1816–92), who became one of the early founders of the electrical industry; also such university-trained pioneers of the modern chemical industry as the Englishman William Perkin (1838–1907) and the Anglo-German Ludwig Mond (1839–1909). But in general, technological invention and the development of industries based on new knowledge were in the hands of craftsmen and artisans with little scientific education but a great deal of mechanical genius. These men considered themselves mechanics and inventors, certainly not engineers or chemists, let alone scientists.

The nineteenth century was also the era of technical-university building. Of the major technical institutions of higher learning only one, the École Polytechnique in Paris, antedates the century; it was founded at the close of the eighteenth century. But by 1901, when the California Institute of Technology in Pasadena admitted its first class, virtually every one of the major technical colleges active in the Western world today had already come into being. Still, in the opening decades of the twentieth century the momentum of technical progress was being carried by the self-taught mechanic without specific technical or scientific education. Neither Henry Ford (1863–1947) nor the Wright brothers (Wilbur, 1867–1912; Orville, 1871–1948) had gone to college.

The technically educated man with the college degree began to assume leadership about the time of World War I, and by the time of the Second World War the change was essentially complete. Technological work since 1940 has been done primarily by men who have been specially educated for such work and who have earned university degrees. Such degrees have almost become prerequisites for technological work. Indeed, since World War II, the men who have built businesses on new technology have as often as not been university professors of physics, chemistry, or engineering, as were most of the men who converted the computer into a salable product.

Technological work has thus become a profession. The inventor has become an engineer, the craftsman a professional. In part this is only a reflection of the uplifting of the whole educational level of the Western world during the last 150 years. The college-trained engineer or chemist in the Western world today is not more educated, considering the relative standard of his society, than the craftsman of 1800 (who, in a largely illiterate society, could read and write). It is our entire society—and not the technologist alone—that has become formally educated and professionalized. But the professionalization of technological work points up the growing complexity of technology and the growth of scientific and technological knowledge. It is proof of a change in attitude toward technology, an acceptance by society, government, education, and business that this work is important, that it requires a thorough grounding in scientific knowledge, and, above all, that it requires many more capable people than "natural genius" could produce.

Technological work has also become increasingly specialized during the twentieth century. Charles Franklin Kettering (1876–1958), the inventive genius of General Motors and for thirty years head of General Motors Research Corporation, represented the nineteenth-century type of inventor, who specialized in invention rather than in electronics, haloid chemistry, or even the automobile. Kettering in 1911 helped invent the electric self-starter, which enabled laymen (and especially lay-women) to drive an automobile. He concluded his long career in the late thirties by converting the clumsy, wasteful, heavy, and inflexible diesel engine into the economical, flexible, and relatively lightweight propulsion unit that has become standard in heavy trucks and railroad locomotives. In between, however, he also developed a nontoxic freezing compound which made household refrigeration possible and, with it, the modern appliance industry; and tetraethyl lead, which, by preventing the "knocking" of internal-combustion engines using high-octane fuel, made possible the high-performance automobile and aircraft engine.

This practice of being an inventor characterized the nineteenth-century technologist altogether. Edison and Siemens in the electrical industry saw themselves as "specialists in invention," as did the father of organic chemistry, Justus von

Liebig (1803–79) of Germany. Even lesser men showed a range of interests and achievements that would seem extraordinary, if not unprofessional, today. George Westinghouse (1846–1914), for instance, took out important patents on a high-speed vertical steam engine; on the generation, transformation, and transmission of alternating current; and on the first effective automatic brake for railroad trains. The German-born Emile Berliner (1851–1929) contributed heavily to early telephone and phonograph technology and also designed one of the earliest helicopter models. And there were others.

This kind of inventor has not yet disappeared—there are men today working as Edison, Siemens, and Liebig worked a century ago. Edwin H. Land (1909–   ), of Polaroid fame, quit college to develop polarizing glass, and has ranged in his work from camera design to missiles, and from optics and the theory of vision to colloidal chemistry. He deliberately describes himself in *Who's Who in America* as an "inventor." But such men who cover the spectrum of applied science and technology are not, as they were in the nineteenth century, the center of technological activity. There we find instead the specialist who works in one increasingly narrow area—electronic circuit design, heat exchange, or high-density polymer chemistry, for instance.

This professionalization and specialization have been made effective by the institutionalization of work in the research laboratory. The research laboratory—and especially the industrial research laboratory—has become the carrier of technological advance in the twentieth century. It is increasingly the research laboratory, rather than the individual, which produces new technology. More and more, technological work is becoming a team effort in which the knowledge of a large number of specialists in the laboratory is brought to bear on a common problem and directed toward a joint technological result.

During the nineteenth century the laboratory was simply the place where work was done that required technical knowledge beyond that of the ordinary mechanic. In industry, testing and plant engineering were the main functions of the laboratory; research was done on the side, if at all. Similarly, the government laboratory during the nineteenth century was essentially a place to test, and all the large government laboratories in the world today (such as the Bureau of Standards in Washington) were founded for that purpose. In the nineteenth-century college or university, the laboratory was used primarily for teaching rather than for research.

Today's research laboratory had its origin in the German organic-chemical industry. The rapid rise of this industry from 1870 on rested squarely on the application of science to industrial production, unheard of until then. However, even those German chemical laboratories were at first given mainly to testing and process engineering, and it was not until 1900 that they were devoted primarily to research. The turning point came with the synthesis of aspirin—the first purely synthetic drug—by Adolf von Baeyer (1835–1917) in 1899. The worldwide success of aspirin within a few years convinced the chemical industry of the value of technological work dedicated to research alone.

Even Edison's famous laboratory in Menlo Park, New Jersey—the most productive research center in the whole history of technological discovery and innovation—was not altogether a modern research laboratory. Although devoted solely to research, as is the modern research laboratory, Menlo Park was still primarily the workshop of a single inventor rather than the team effort that characterizes the industrial or university research laboratory of today. Many of Edison's assistants became successful inventors in their own right, for instance, Frank J. Sprague (1857–1934), who developed the first practical electric streetcar. But these men became productive technologists only after they had left Menlo Park and Edison's employ. While there, they were just the great man's helpers.

After the turn of the century, new research laboratories suddenly appeared on both sides of the Atlantic. The German chemical industry rapidly built great laboratories that helped to give Germany a worldwide monopoly on dyestuffs, pharmaceuticals, and other organic chemicals before World War I. In Germany, too, shortly after 1900, were founded the big governmental research laboratories of the Kaiser Wilhelm Society (now the Max Planck Society), where senior scientists and scientific teams, free from all teaching obligations, could engage in research alone. On this side of the Atlantic C. P. Steinmetz (1865–1923) began, at about the same time, to build the first modern research laboratory in the electrical industry, the great research center of the General Electric Company in Schenectady. Steinmetz understood, perhaps even better than the Germans, what he was doing, and the pattern he laid down for the General Electric Research Laboratory is by and large that followed by all major industrial and governmental research centers to this day.

The essence of the modern research laboratory is not its size. There are some very large laboratories, working for governments or for large companies, and also numerous small research laboratories, many employing fewer technologists and scientists than did some nineteenth-century establishments; and there is no apparent relationship between the size of the research laboratory and its results. What distinguishes today's research laboratory from any predecessor is, first, its exclusive interest in research, discovery, and innovation. Second, the research laboratory brings together men from a wide area of disciplines, each contributing his specialized knowledge. Finally, the research laboratory embodies a new methodology of technological work squarely based on the systematic application of science to technology.

It is a great strength of the research laboratory that it can be both "specialist" and "generalist," permitting an individual to work alone or a team to work together. Quite a few Nobel Prize winners have done their research work in industrial laboratories such as those of the Bell Telephone System or the General Electric Company. Similarly nylon (1937), one of the first building blocks of today's plastic industry, was developed by W. H. Carothers (1896–1937) working by himself in the DuPont laboratory during the thirties. The research laboratory provides an individual with access to skills and facilities which greatly increase his capacity. It can at the same time, however, organize a team effort for a specific task and thus create a collective generalist with a greater range of skills and knowledge than any individual, no matter how gifted, could possibly acquire in one lifetime.

Before World War I the research laboratory was still quite rare. Between World War I and World War II it became standard in a number of industries, primarily the chemical, pharmaceutical, electrical, and electronics industries. Since World War II, research activity has become as much of a necessity in industry as a manufacturing plant, and as central in its field as is the infantry soldier for defense, or the trained nurse in medicine.

The Methods of Technological Work

Hand in hand with changes in the structure of technological work go changes in the basic approach to and methods of work. Technology has become science-based. Its method is now "systematic research." And what was formerly "invention" is "innovation" today.

Historically the relationship between science and technology has been a complex one, and it has by no means been thoroughly explored nor is it truly understood as yet. But it is certain that the scientist, until the end of the nineteenth century, with rare exceptions, concerned himself little with the application of his new scientific knowledge and even less with the technological work needed to make knowledge applicable. Similarly, the technologist, until recently, seldom had direct or frequent contact with the scientist and did not consider his findings of primary importance to technological work. Science required, of course, its own technology—a very advanced technology at that, since all along the progress of science has depended upon the development of scientific instruments. But the technological advances made by the scientific instrument maker were not, as a rule, extended to other areas and did not lead to new products for the consumer or to new processes for artisan and industry. The first instrument maker to become important outside of the scientific field was James Watt, the inventor of the steam engine.

Not until almost seventy-five years later, that is until 1850 or so, did scientists themselves become interested in the technological development and application of their discoveries. The first scientist to become a major figure in technology was Justus von Liebig, who in the mid-nineteenth century developed the first synthetic fertilizer and also a meat extract (still sold all over Europe under his name) which was, until the coming of refrigeration in the 1880's, the only way to store and transport animal proteins. In 1856 Sir William H. Perkin in England isolated, almost by accident, the first aniline dye and immediately built a chemical business on his discovery. Since then, technological work in the organic-chemicals industry has tended to be science-based.

About 1850 science began to affect another new technology—electrical engineering. The great physicists who contributed scientific knowledge of electricity during the century were not themselves engaged in applying this knowledge to products and processes; but the major nineteenth-century technologists of electricity closely followed the work of the scientists. Siemens and Edison were thoroughly familiar with the work of physicists such as Michael Faraday (1791–1867) and Joseph Henry (1791–1878). And Alexander Graham Bell (1847–1927) was led to his work on the telephone through the researches of Hermann von Helmholtz (1821–94) on the reproduction of sound. Guglielmo Marconi (1874–1910) developed radio on the foundation Heinrich Hertz (1857–94) had laid with his experimental confirmation of Maxwell's electromagnetic-wave propagation theory; and so on. From its beginnings, therefore, electrical technology has been closely related to the physical science of electricity.

Generally, however, the relationship between scientific work and its technological application, which we today take for granted, did not begin until after the turn of the twentieth century. As previously mentioned, such typically modern devices as the automobile and the airplane benefitted little from purely theoretical scientific work in their formative years. It was World War I that brought about the change: in all belligerent countries scientists were mobilized for the war effort, and it was then that industry discovered the tremendous power of science to spark technological ideas and to indicate technological solutions. It was at that time also that scientists discovered the challenge of technological problems.

Today technological work is, for the most part, consciously based on scientific effort. Indeed, a great many industrial research laboratories do work in "pure" research, that is, work concerned exclusively with new theoretical knowledge rather than with the application of knowledge.

And it is a rare laboratory that starts a new technological project without a study of scientific knowledge, even where it does not seek new knowledge for its own sake. At the same time, the results of scientific inquiry into the properties of nature—whether in physics, chemistry, biology, geology, or another science—are immediately analyzed by thousands of "applied scientists" and technologists for their possible application to technology.

Technology is not, then, "the application of science to products and processes," as is often asserted. At best this is a gross oversimplification. In some areas—for example, polymer chemistry, pharmaceuticals, atomic energy, space exploration, and computers—the line between "scientific inquiry" and "technology" is a blurred one; the scientist who finds new basic knowledge and the technologist who develops specific processes and products are one and the same man. In other areas, however, highly productive efforts are still primarily concerned with purely technological problems, and have little connection to science as such. In the design of mechanical equipment—machine tools, textile machinery, printing presses, and so forth—scientific discoveries as a rule play a very small part, and scientists are not commonly found in the research laboratory. More important is the fact that science, even where most relevant, provides only the starting point for technological efforts. The greatest amount of work on new products and processes comes well *after* the scientific contribution has been made. "Know-how," the technologist's contribution, takes a good deal more time and effort in most cases than the scientist's "know-what"; but though science is no substitute for today's technology, it is the basis and starting point.

While we know today that our technology is based on science, few people (other than the technologists themselves) realize that technology has become in this century somewhat of a science in its own right. It has become research—a separate discipline having its own specific methods.

Nineteenth-century technology was "invention"—not managed or organized or systematic. It was, as our patent laws, now two hundred years old, still define it, "flash of insight." Of course hard work, sometimes for decades, was usually required to convert this "flash" into something that worked and could be used. But nobody knew how this work should be done, how it might be organized, or what one could expect from it. The turning point was probably Edison's work on the electric light bulb in 1879. As his biographer Matthew Josephson points out, Edison did not intend to do organized research. He was led to it by his failure to develop through "flash of genius" a workable electric light. This forced him, very much against his will, to work through the specifications of the solution needed, to spell out in considerable detail the major steps that had to be taken, and then to test systematically one thousand six hundred different materials to find one that could be used as the incandescent element for the light bulb he sought to develop. Indeed, Edison found that he had to break through on three major technological fronts at once in order to have domestic electric lighting. He needed an electrical energy source producing a well-regulated voltage of essentially constant magnitude; a high vacuum in a small glass container; and a filament that would glow without immediately burning up. And the job that Edison expected to finish by himself in a few weeks required a full year and the work of a large number of highly trained assistants, that is, a research team.

There have been many refinements in the research method since Edison's experiments. Instead of testing one thousand six hundred different materials, we would today, in all probability, use conceptual and mathematical analysis to narrow the choices considerably (this does not always work, however; current cancer research, for instance, is testing more than sixty thousand chemical substances for possible therapeutic action). Perhaps the greatest improvements have been in the management of the research team. There was, in 1879, no precedent for such a team effort, and Edison had to improvise research management as he went along. Nevertheless, he clearly saw the basic elements of research discipline: (1) a definition of the need—for Edison, a reliable and economical system of converting electricity into light; (2) a clear goal—a transparent container in which resistance to a current would heat up a substance to white heat; (3) identification of the major steps to be taken and the major pieces of work that had to be done—in his case, the power source, the container, and the filament; (4) constant feedback from the results of the work on the plan; for example, Edison's finding that he needed a high vacuum rather than an inert gas as the environment for his filament made him at once change the direction of research on the container; and finally (5) organization of the work so that each major segment is assigned to a specific work team.

These steps together constitute to this day the basic method and the system of technological work. October 21, 1879, the day on which Edison first had a light bulb that would burn for more than a very short time, therefore, is not only the birthday of electric light; it marks the birth of modern technological research as well. Yet whether Edison himself fully understood what he had accomplished is not clear, and certainly few people at the time recognized that he had found a generally applicable method of technological and scientific inquiry. It took twenty years before Edison was widely imitated, by German chemists and bacteriologists in their laboratories and in the General Electric laboratory in the United States. Since then, however, technological work has progressively developed as a discipline of methodical inquiry everywhere in the Western world.

Technological research has not only a different methodology from invention; it leads to a different approach, known as innovation, or the purposeful and deliberate attempt to bring about, through technological means, a distinct change in the way man lives and in his environment—the economy, the society, the community, and so on. Innovation may begin by defining a need or an opportunity, which then leads to organizing technological efforts to find a way to meet the need or exploit the opportunity. To reach the moon, for instance, requires a great deal of new technology; once the need has been defined, technological work can be organized systematically to produce this new technology. Or innovation can proceed from new scientific knowledge and an analysis of the opportunities it might be capable of creating. Plastic fibers, such as nylon, came into being in the 1930's as a result of systematic study of the opportunities offered by the new understanding of polymers (that is, long chains of organic molecules), which chemical scientists (mostly in Germany) had gained during World War I.

Innovation is not a product of the twentieth century; both Siemens and Edison were innovators as much as inventors. Both started out with the opportunity of creating big new industries—the electric railway (Siemens), and the electric lighting industry (Edison). Both men analyzed what new technology was needed and went to work creating it. Yet only in this century—and largely through the research laboratory and its approach to research—has innovation become central to technological effort.

In innovation, technology is used as a means to bring about change in the economy, in society, in education, in warfare, and so on. This has tremendously increased the impact of technology. It has become the battering ram which breaks through even the stoutest ramparts of tradition and habit. Thus modern technology influences traditional society and culture in underdeveloped countries. But innovation means also that technological work is not done only for technological reasons but for the sake of a nontechnological economic, social, or military end.

Scientific discovery has always been measured by what it adds to our understanding of natural phenomena. The test of invention is, however, technical—what new capacity it gives us to do a specific task. But the test of innovation is its impact on the way people live. Very powerful innovations may, therefore, be brought about with relatively little in the way of new technological invention.

A very good example is the first major innovation of the twentieth century, mass production, initiated by Henry Ford between 1905 and 1910 to produce the Model T automobile. It is correct, as has often been pointed out, that Ford contributed no important technological invention. The mass-production plant, as he designed and built it between 1905 and 1910, contained not a single new element: interchangeable parts had been known since before Eli Whitney, a century earlier; the conveyor belt and other means of moving materials had been in use for thirty years or more, especially in the meat-packing plants of Chicago. Only a few years before Ford, Otto Doering, in building the first large mail-order plant in Chicago for Sears, Roebuck, used practically every one of the technical devices Ford was to use at Highland Park, Detroit, to turn out the Model T. Henry Ford was himself a highly gifted inventor who found simple and elegant solutions to a host of technical problems—from developing new alloy steels to improving almost every machine tool used in the plant. But his contribution was an innovation: a technical solution to the economic problem of producing the largest number of finished products with the greatest reliability of quality at the lowest possible cost. And this innovation has had greater impact on the way men live than many of the great technical inventions of the past.

The Systems Approach

Mass production exemplifies, too, a new dimension that has been added to technology in this century: the systems approach. Mass production is not a thing, or a collection of things; it is a concept—a unified view of the productive process. It requires, of course, a large number of "things," such as machines and tools. But it does not start with them; they follow from the vision of the system.

The space program today is another such system, and its conceptual foundation is genuine innovation. Unlike mass production, the space program requires a tremendous amount of new invention, as well as new scientific discovery. Yet the fundamental scientific concepts underlying it are not at all new—they are, by and large, Newtonian physics. What is new is the idea of putting men into space by a systematic, organized approach.

Automation is a systems concept, closer to Ford's mass production than to the space program. There had been examples of genuine automation long before anyone coined the term. Every oil refinery built in the past forty years has been essentially automated. But not until someone saw the entire productive process as one continuous, controlled flow of materials did we see automation. This has led to a tremendous amount of new technological activity to develop computers, process controls for machines, materials-moving equipment, and so on. Yet the basic technology to automate a great many industrial processes had been present for a long time, and all that was lacking was the systems approach to convert them to the innovation of automation.

The systems approach, which sees a host of formerly unrelated activities and processes as all parts of a larger, integrated whole, is not something technological in itself. It is, rather, a way of looking at the world and at ourselves. It owes much to *Gestalt* psychology (from the German word for "configuration" or "structure"), which demonstrated that we do not see lines and points in a painting but configurations—that is, a whole—and that we do not hear individual sounds in a tune but only the tune itself—the configuration. And the systems approach was also generated by twentieth-century trends in technology: the linking of technology and science, the development of the systematic discipline of research, and innovation. The systems approach is, in fact, a measure of our newly found technological capacity. Earlier ages could visualize systems but they lacked the technological means to realize such visions.

The systems approach also tremendously increases the power of technology. It permits today's technologists to speak of materials rather than of steel, glass, paper, or concrete, each of which has, of course, its own (very old) technology. Today we see a generic concept—materials—all of which are arrangements of the same fundamental building blocks of matter. Thus it happens that we are busy designing materials without precedent in nature: synthetic fibers, plastics, glass that does not break and glass that conducts electricity, and so on. We increasingly decide first what end use we want and then choose or fashion the material we want to use. We define, for example, the specific properties we want in a container and then decide whether glass, steel, aluminum, paper, one of a host of plastics, or any one of hundreds of materials in combination will be the best material for it. This is what is meant by a "materials revolution" whose specific manifestations are technological, but whose roots are to be found in the systems concept.

We are similarly on the threshold of an "energy revolution"—making new use of such sources of energy as atomic reaction, solar energy, the tides, and so forth; but also with a new systems concept: energy. Again this concept is the result of major technological developments—especially, of course, in atomic power—and the starting point for major new technological work. Ahead of us, and barely started, is the greatest systems job we can see now: the systematic exploration and development of the oceans.

Water covers far more of the earth's surface than does land. And since water, unlike soil, is penetrated by the rays of the sun for a considerable depth, the life-giving process of photosynthesis covers infinitely more area in the seas than it does on land—apart from the fact that every single square inch of the ocean is fertile. And the sea itself, as well as its bottom, contains untold riches in metals and minerals. Yet, even today, on the oceans man is still a hunter and a nomad rather than a cultivator. He is in the same early stage of development as our ancestors almost ten thousand years ago when they first tilled the soil. Comparatively minor efforts to gain knowledge of the oceans and to develop technology to cultivate them should therefore yield returns—not only in knowledge, but in food, energy, and raw materials also—far greater than anything we could get from exploiting the already well-explored lands of the continents. Oceanic development, rather than space exploration, might well turn out to be the real frontier in the next century. Underlying this development will be the concept of the oceans as a system, resulting from such technological developments as the submarine, and in turn sparking such new technological efforts as the Mohole project to drill through the earth's hard crust beneath the ocean.

There are many other areas where the systems approach is likely to have a profound impact, where it may lead to major technological efforts, and through them, to major changes in the way we live and in our capacity to do things. One such example is the modern city—itself largely a creation of modern technology.

One of the greatest nineteenth-century inventions was invention itself, as has been said many times. It underlay the explosive technological development of the years between 1860 and 1900, "the heroic age of invention." It might similarly be said that the great invention of the early twentieth century was innovation: it underlies the deliberate attempt to organize purposeful changes of whole areas of life which characterizes the systems approach.

Innovation and the systems approach are only just emerging. Their full impact is almost certainly still ahead. But they are already changing man's life, society, and his world view. And they are profoundly changing technology itself and its role.

5 | Technology and Society in the Twentieth Century

The Pretechnological Civilization of 1900

Modern man everywhere takes technological civilization for granted. Even primitive people in the jungles of Borneo or in the High Andes, who may themselves still live in the Early Bronze Age and in mud huts as they have for thousands of years, need no explanation when the movie they are watching shows the flipping of a light switch, the lifting of a telephone receiver, the starting of an automobile or plane, or the launching of another satellite. In mid-twentieth century the human race has come to feel that modern technology holds the promise of conquering poverty on the earth and of conquering outer space beyond. We have learned, too, that it carries the threat of snuffing out all humanity in one huge catastrophe. Technology stands today at the very center of human perception and human experience.

On the other hand, at the beginning of the twentieth century modern technology barely existed for the great majority of people. In terms of geography, the Industrial

First published in vol. 2 of *Technology in Western Civilization,* ed. Melvin Kranzberg and Carroll W. Pursell, Jr. (New York: Oxford University Press, 1967).

Revolution and its technological fruits were largely confined, in 1900, to the small minority of mankind that is of European descent and lives around the North Atlantic shores. Only Japan, of the non-European, non-Western countries, had then begun to build up a modern industry and modern technology, and in 1900 modern Japan was still in its infancy. In Indian village, Chinese town, and Persian bazaar, life was still preindustrial, still untouched by the steam engine and telegraph, and by all other new tools of the West. It was, indeed, almost an axiom—for Westerner and non-Westerner alike—that modern technology was, for better or worse, the birthright of the white man and restricted to him. This assumption underlay the imperialism of the period before World War I, and it was shared by such eminent non-Westerners as Rabindranath Tagore (1861–1941), the Nobel Prize-winning Indian poet, and Mahatma Gandhi (1869–1948), who, just before World War I, began his long fight for Indian independence. There was, indeed, enough apparent factual support for this belief to survive, if only as a prejudice, until World War II. Hitler, for instance, made the Japanese "honorary Aryans" and considered them "Europeans in disguise" primarily because they had mastered modern technology. And in the United States the myth lingered on in the widespread belief, before Pearl Harbor, that the Japanese, not being of European stock, were not proficient in handling such weapons of modern technology as planes or battleships.

Yet, in the West, indeed even in the most highly developed countries—England, the United States, and Germany —modern technology played in 1900 only a minor role in the lives of most people, the majority of whom were then still farmers or artisans living either in the countryside or in small towns. The tools they used and the life they led were preindustrial, and they remained unaware of the modern technology that was so rapidly growing up all around them.

Only in a few large cities had modern technology imposed itself upon daily life—in the street railways, increasingly powered by electricity after 1890, and in the daily paper, dependent upon the telegraph and printed on steam-driven presses. Only there had modern technology crossed the threshold of the home with electric lights and the telephone.

Even so, to Western man in 1900, modern technology had become tremendously exciting. It was the time of the great international exhibitions in every one of which a new "miracle" of technical invention stole the show. These were also the years in which technological fiction became a best seller from Moscow to San Francisco. About 1880, books by the Frenchman Jules Verne (1828–1905), such as *Journey to the Center of the Earth* and *Twenty Thousand Leagues under the Sea,* became wildly popular. By 1900 the English novelist H. G. Wells (1866–1946), whose works included the technological romance *The Time Machine* (1893), had become more popular still. And there was virtually unbounded faith in the benevolence of technological progress. All this excitement was, however, focused on *things.* That these things could and would have an impact on society and on the way people behaved and thought had not occurred to many.

The advances in technology in this century are, indeed, awe-inspiring. Nevertheless, it can be argued that the foundations for most of them had been well laid by 1900, and certainly by 1910. The electric light, the telephone, the moving picture, the phonograph, and the automobile had all been invented by 1900 and were, indeed, being sold aggressively by prosperous and growing companies. And the airplane, the vacuum tube, and radio telegraphy were invented in the opening years of the new century.

The changes technology has wrought in society and culture since then could, however, not have been seen by the men of 1900. The geographical explosion of technology has created the first worldwide civilization; and it is a technological civilization. It has already shifted the power center of the world away from Western Europe thousands of miles to both West and East. More important still, modern technology in this century has made men reconsider old concepts, such as the position of women in society, and it has remade basic institutions—work, education, and warfare, for example. It has shifted a large number of people in the technologically advanced countries from working with their hands to working, almost without direct contact with materials and tools, with their minds. It has changed the physical environment of man from one of nature to the man-made big city. It has further changed man's horizon. While it converts the entire world into one rather tight community sharing knowledge, information, hopes, and fears, technology has brought outer space into man's immediate, conscious experience. It has converted an apocalyptic promise and an apocalyptic threat into concrete possibilities here and now: offering both the utopia of a world without poverty and the threat of the final destruction of humanity.

Finally, in the past sixty years man's view of technology itself has changed. We no longer see it as concerned with *things* only; today it is a concern of man as well. As a result of this new perspective we have come to realize that technology is not, as our grandparents believed, the magic wand that can make all human problems and limitations disappear. We now know that technological potential is, indeed, even greater than they thought. But we have also learned that technology, as a creature of man, is as problematical, as ambivalent, and as capable of good or evil, as is its creator.

This paper will attempt to point out some of the most important changes which modern technology has brought about in society and culture, and some changes in our own view of and approach to technology thus far, in the twentieth century.

Technology Remakes Social Institutions

Twentieth-century history, up to the 1960's, can be divided into three major periods: the period before the outbreak of the First World War in 1914—a period culturally and politically much like the nineteenth century; the First World War and the twenty years from 1918 to the outbreak of World War II in 1939; and from World War II until today. In each of these periods modern technology has shaped basic institutions of Western society. And in the most recent period it has started to undermine and remake, also, many of the basic institutions of non-Western society.

Emancipation of Women

In the years before World War I technology, in large measure, brought about the emancipation of women and gave them a new position in society. No nineteenth-century feminist, such as Susan B. Anthony, had as strong an impact on the social position of women as did the typewriter and telephone. If the "Help Wanted" advertisement of 1880 said "stenographer" or "telegrapher," everybody knew that a man was wanted, whereas the ad of 1910 for a stenographer or telephone operator was clearly offering a woman's job. The typewriter and the telephone enabled the girl from a decent family to leave home and make a respectable living on her own, not dependent on a husband or father. The need for women to operate typewriters and switchboards forced even the most reluctant European governments to provide public secondary education for girls, the biggest single step toward granting women equality. The flood of respectable and well-educated young women in offices then made heavy demands for changes in the old laws that withheld from women the right to enter into contracts or to control their own earnings and property, and finally forced men by 1920 to give women the vote almost everywhere in the Western world.

Changes in the Organization of Work

Technology soon began to bring about an even greater transformation about the time of World War I. It started to make over the manual work that had always provided a livelihood for the great majority of people—as it still does in technologically underdeveloped countries. The starting point was the application of modern technological principles to manual work, which went under the name of Scientific Management and was largely developed by an American, Frederick Winslow Taylor (1856–1915).

While Henry Ford made the systems innovation of mass production, Taylor applied to manual operations the principles that machine designers during the nineteenth century had learned to apply to the work of a tool; he identified the work to be done; broke it down into its individual operations; designed the right way to do each operation; and finally he put the operations together, this time in the sequence in which they could be done fastest and most economically. All this strikes us today as commonplace, but it was the first time that work had been looked at; throughout history it had always been taken for granted.

The immediate result of Scientific Management was a revolutionary cut in the costs of manufactured goods—often to one-tenth, sometimes to one-twentieth of what they had been before. What had been rare luxuries inaccessible to all but the rich, such as automobiles or household appliances, rapidly became available to the broad masses. More important, perhaps, is the fact that Scientific Management made possible sharp increases in wages while at the same time lowering the total cost of the product. Hitherto lower costs of a finished product had always meant lower wages to the worker producing it. Scientific Management preached the contrary: that lower costs should mean higher wages and higher income for the worker. To bring this about was indeed Taylor's main intent and that of his disciples, who, unlike many earlier technologists, were motivated as much by social as by technical considerations. "Productivity" at once became something the technologist could raise, if not create. And with it, the standard of living of a whole economy might be raised, something totally impossible—indeed, almost unimaginable—at any earlier time.

At the same time, Scientific Management rapidly changed the structure and composition of the work force. It first led to wholesale upgrading of the labor force. The unskilled "laborer" working at a subsistence wage, who constituted the largest single group in the nineteenth-century labor force, became obsolete. In his place appeared a new group, the machine operators—the men on the automobile assembly line, for instance. They themselves were no more skilled, perhaps, than the laborers, but the technologist's knowledge had been injected into their work through Scientific Management so that they could be paid—and were soon being paid—the wages of highly skilled workers. Between 1910 and 1940 the machine operators became the largest single occupational group in every industrial country, pushing both farmers and laborers out of first place. The consequences for mass consumption, labor relations, and politics were profound and are still with us.

Taylor's work rested on the assumption that knowledge, rather than manual skill, was the fundamental productive resource. Taylor himself preached that productivity required that "doing" be divorced from "planning," that is, that it be based on systematic technological knowledge. His work resulted in a tremendous expansion of the number of educated people needed in the work force and, ultimately, in a complete shift in the focus of work from labor to knowledge.

What is today called automation is conceptually a logical extension of Taylor's Scientific Management. Once operations have been analyzed as if they were machine operations and organized as such (and Scientific Management did this successfully), they should be capable of being performed by machines rather than by hand. Taylor's work immediately increased the demand for educated people in the work force, and eventually, after World War II, it began to produce a work force in advanced countries like the United States in which educated people applying their knowledge to the job are the actual "workers," and outnumber the manual workers, whether laborers, machine operators, or craftsmen.

The substitution of knowledge for manual effort as the productive resource in work is the greatest change in the history of work, which is, of course, a process as old as man himself. This change is still in progress, but in the industrially advanced countries, especially in the United States, it has already completely changed society. In 1900 eighteen out of every twenty Americans earned their living by working with their hands, ten of the eighteen as farmers. By 1965, only five out of twenty of a vastly increased American labor force did manual work, and only one worked on the farm. The rest earned their living primarily with knowledge, concepts, or ideas—altogether, with things learned in school rather than at the workbench. Not all of this knowledge is, of course, advanced; the cashier in the cafeteria is also a "knowledge worker," though of very limited extent. But all of it is work that requires education, that is, systematic mental training rather than skill in the sense of exposure to experience.

The Role of Education

As a result, the role of education in twentieth-century industrial society has changed—another of the very big changes produced by technology. By 1900 technology had advanced so far that literacy had become a social need in the industrial countries. A hundred years earlier literacy was essentially a luxury as far as society was concerned; only a handful of people—ministers, lawyers, doctors, government officials, and merchants—needed to be able to read and write. Even for a high-ranking general, such as Wellington's partner at Waterloo, the Prussian Field Marshal Bluecher, illiteracy was neither a handicap nor a disgrace. In the factory or the business office of 1900, however, one had to be able to read and write, if only at an elementary school level. By 1965 those without a substantial degree of higher education, more advanced than anything that had been available even to the most educated two hundred years ago, were actually becoming unemployable. Education has moved, from having been an ornament, if not a luxury, to becoming the central economic resource of technological society. Education is therefore rapidly becoming a center of spending and investment in the industrially developed society.

This stress on education is creating a changed society; access to education is best given to everyone, if only because society needs all the educated people it can get. The educated man resents class and income barriers which prevent the full exercise of this knowledge, and because society requires and values the services of the expert it must allow him full recognition and rewards for his talents. In a completely technological civilization, education replaces money and rank as the index of status and opportunities.

Change in Warfare

By the end of World War II technology had completely changed the nature of warfare, and altered the character of war as an institution. When Karl von Clausewitz (1780–1831), the father of modern strategic thought, called war "a continuation of policy by other means" he only expressed in an epigram what every politician and every military leader had known all along. War was always a gamble. War was cruel and destructive. War, the great religious leaders always preached, is sin. But war was also a normal institution of human society and a rational tool of policy. Many of the contemporaries, including Clausewitz himself, considered Napoleon wicked, but none thought him insane for using war to impose his political will on Europe.

The dropping of the first atomic bomb on Hiroshima in 1945 changed all this. Since then it has become increasingly clear that major war no longer can be considered normal, let alone rational. Total war has ceased to be a usable institution of human society because in full-scale, modern technological warfare, there is no defeat, just as there is no victory. There is only total destruction. There are no neutrals and no noncombatants, for the risk of destruction extends to the entire human race.

A Worldwide Technological Civilization

After World War II modern technology established a worldwide technological civilization. Modern inventions had, of course, been penetrating the non-Western world steadily since 1900: bicycle, automobile, truck, electric lights, telephone, phonograph, movie, and radio, and so on.

In most areas, these remained culturally surface phenomena until the 1940's. The Bedouin on the Arabian Desert took to carrying a battery radio on the camel; but he used it mainly to receive the muezzin's call to evening prayer direct from Mecca. World War II brought modern technology in its most advanced forms directly to the most remote corners of the earth. The airplane became as familiar as the camel had been. All armies required modern technology to provide the sinews of war and the instruments of warfare. And all used non-Western people either as soldiers in technological war or as workers on modern machinery to provide war material. This made everyone in the world aware of the awesome power of modern technology.

This, however, might not have had a revolutionary impact upon older, non-Western, nontechnological societies but for the promise of Scientific Management to make possible systematic economic development. The new-found power to create productivity through the systematic effort we now call industrialization has raised what President John F. Kennedy called "the rising tide of human expectations," the hope that technology can banish the age-old curse of disease and early death, of grinding poverty, and ceaseless toil. And whatever else this may require, it demands acceptance by society of a thoroughly technological civilization.

The shift of focus in the struggle between social ideologies shows this clearly. Before World War II free enterprise and communism were generally measured throughout the world by their respective claims to have superior ability to create a free and just society. Since World War II the question has largely been: which system is better at speeding economic development to a modern technological civilization? India offers another illustration. Until his death in 1948, Mahatma Gandhi opposed industrialization and sought a return to a pre-industrial technology, symbolized in the hand spinning wheel. His close comrade and disciple Jawaharlal Nehru

(1889–1964) was forced, however, by public opinion to embrace "economic development," that is, forced-draft industrialization emphasizing the most modern technology, as soon as he became the first prime minister of independent India in 1947.

Even in the West, where it grew out of the indigenous culture, technology has in the twentieth century raised fundamental problems for society and has challenged—if not overthrown—deeply rooted social and political institutions. Wherever technology moves it has an impact on the position of women in society; on work and the worker; on education and social mobility; and on warfare. Since this is the case, in the non-Western societies modern technology demands a radical break with social and cultural tradition; and it produces a fundamental crisis in society. How the non-Western world will meet this crisis will, in large measure, determine what man's history will be in the latter twentieth century— even, perhaps, whether there will be any more human history. But unless the outcome is the disappearance of man from this planet, our civilization will remain irrevocably a common technological civilization.

Man Moves into a Man-made Environment

By 1965 the number living on the land and making their living off it had dwindled in the U.S. to one out of every twenty. Man had become a city-dweller. At the same time, man in the city increasingly works with his mind, removed from materials. Man in the twentieth century has thus moved from an environment that was essentially still nature to an environment, the large city and knowledge work, that is increasingly man-made. The agent of this change has, of course, been technology.

Technology, as has been said before, underlies the shift

from manual to mental work. It underlies the tremendous increase in the productivity of agriculture which, in technologically developed countries like the United States or those of Western Europe, has made one farmer capable of producing, on less land, about fifteen times as much as his ancestor did in 1800 and almost ten times as much as his ancestors in 1900. It therefore enabled man to tear himself away from his roots in the land to become a city-dweller.

Indeed, urbanization has come to be considered the index of economic and social development. In the United States and in the most highly industrialized countries of Western Europe up to three-quarters of the population now live in large cities and their suburbs. A country like the Soviet Union, that still requires half its people to work on the land to be adequately fed, is, no matter how well developed industrially, an "underdeveloped country."

The big city is, however, not only the center of modern technology; it is also one of its creations. The shift from animal to mechanical power, and especially to electrical energy (which needs no pasture lands), made possible the concentration of large productive facilities in one area. Modern materials and construction methods make it possible to house, move, and supply a large population in a small area. Perhaps the most important prerequisite of the large modern city, however, is modern communications, the nerve center of the city and the major reason for its existence. The change in the type of work a technological society requires is another reason for the rapid growth of the giant metropolis. A modern society requires that an almost infinite number of specialists in diverse fields of knowledge be easy to find, easily accessible, and quickly and economically available for new and changing work. Businesses or government offices move to the city, where they can find the lawyers, accountants, advertising men, artists, engineers, doctors, scientists, and other trained personnel they need. Such knowledgeable people, in turn, move to the big city to have easy access to their potential employers and clients.

Only sixty years ago, men depended on nature and were primarily threatened by natural catastrophes, storms, floods, or earthquakes. Men today depend on technology, and our major threats are technological breakdowns. The largest cities in the world would become uninhabitable in forty-eight hours were the water supply or the sewage systems to give out. Men, now city-dwellers, have become increasingly dependent upon technology, and our habitat is no longer a natural ecology of wind and weather, soil and forest, but a man-made ecology. Nature is no longer an immediate experience; New York City children go to the Bronx Zoo to see a cow. And whereas sixty years ago a rare treat for most Americans was a trip to the nearest market town, today most people in the technologically advanced countries attempt to "get back to nature" for a vacation.

Modern Technology and the Human Horizon

Old wisdom—old long before the Greeks—held that a community was limited to the area through which news could easily travel from sunrise to sunset. This gave a "community" a diameter of some fifty miles or so. Though each empire—Persian, Roman, Chinese, and Inca—tried hard to extend this distance by building elaborate roads and organizing special speedy courier services, the limits on man's horizon until the late nineteenth century remained unchanged and confined to how far one man could travel by foot or on horseback in one day.

By 1900 there were already significant changes. The railroad had extended the limit of one day's travel to seven hundred miles or more—the distance from New York to Chicago or from Paris to Berlin. And, for the first time, news .

and information were made independent of the human carrier through the telegraph, which carried them everywhere practically instantaneously. It is no accident that a very popular book of technological fiction to this day is Jules Verne's *Around the World in Eighty Days*. For the victory of technology over distance is, perhaps, the most significant of all the gifts modern technology has brought man.

Today the whole earth has become a local community if measured by the old yardstick of one day's travel. The commercial jet plane can reach, in less than twenty-four hours, practically any airport on earth. And unlike any earlier age, the common man can and does move around and is no longer rooted to the small valley of his birth. The motor vehicle has given almost everyone the power of mobility, and with physical ability to move around comes a new mental outlook and a new social mobility. The technological revolution on the American farm began in earnest when the farmer acquired wheels; he immediately became mobile, too, in his mental habits and accessible to new ideas and techniques. The beginning of the Negro drive for civil rights in the American South came with the used car. Behind the wheel of a Model T a Negro was as powerful as any white man, and his equal. Similarly, the Indian worker on the Peruvian sugar plantation who has learned to drive a truck will never again be fully subservient to the white manager. He has tasted the new power of mobility, a greater power than the mightiest kings of yesterday could imagine. It is no accident that young people everywhere dream of a car of their own; four-wheeled mobility is a true symbol of freedom from the restraints of traditional authority.

News, data, information, and pictures have become even more mobile than people. They travel in "real time," that is, they arrive at virtually the same time as they happen. They have, moreover, become universally accessible. The radio brings to anyone in possession of a cheap and simple receiving set news in his own language from any of the world's major capitals. Television and movies present the world everywhere as immediate experience. And beyond the earth itself the horizon of man has, within the last two decades, extended out into stellar space. It is not just a bigger world, therefore, into which twentieth-century technology has moved the human being; it is a different world.

Technology and Man

In this different world, technology itself is seen differently; we are aware of it as a major element in our lives, indeed, in human life throughout history. We are becoming aware that the major questions regarding technology are not technical but human questions, and are coming to understand that a knowledge of the history and evolution of technology is essential to an understanding of human history. Furthermore, we are rapidly learning that we must understand the history, the development, and the dynamics of technology in order to master our contemporary technological civilization, and that, unless we do so, we will have to submit to technology as our master.

The naive optimism of 1900, which expected technology somehow to create paradise on earth, would be shared by few people today. Most would also ask: What does technology do *to* man as well as for him? For it is only too obvious that technology brings problems, disturbances, and dangers as well as benefits. First, economic development based on technology carries the promise of abolishing want over large parts of the earth. But there is also the danger of total war leading to total destruction; and the only way we now know to control this danger is by maintaining in peacetime a higher level of armaments in all major industrial countries than any nation has ever maintained. This hardly seems a fitting answer to the problem, let alone a permanent one. Also, modern public-health technology—insecticides above all—has everywhere greatly increased man's life span. But since birth rates in the underdeveloped countries remain at their former high level while the death rates have declined, the world's poor nations are threatened by a population explosion which not only eats up all the fruits of economic development but threatens world famine and new pestilence. In government, modern technology and the modern economy founded on it have outmoded the national state as a viable unit. Even Great Britain, with fifty million inhabitants, has been proven in recent decades to have too small a productive base and market for independent economic survival and success. Nationalism is still the most potent political force, as developments in the new nations of Asia and Africa clearly show; yet the revolution in transportation and communication has made national borders an anachronism, respected by neither airplanes nor electronic waves.

The metropolis has become the habitat of modern man. Yet paradoxically we do not know how to make it habitable. We have no effective political institutions to govern it. Urban decay and traffic jams, overcrowding and urban crime, juvenile delinquency and loneliness are endemic in all modern great cities. No one looking at any of the world's big cities would maintain that they offer an aesthetically satisfying environment. The divorce from direct contact with nature in work with soil and materials has permitted us to live much better; yet technological change itself seems to have speeded up so much as to deprive us of the psychological and cultural bearings we need.

The critics of technology and dissenters from technological optimism in 1900 were lonely voices. Disenchantment with technology did not set in until after World War I and the coming of the Great Depression. The new note was first fully struck in the novel *Brave New World* by the English writer Aldous Huxley (1894–1964), published in 1932, at the very bottom of the Depression. In this book Huxley portrayed the society of the near future as one in which technology had become the master, and man, its abject slave, was kept in bodily comfort without knowledge of want or pain but also without freedom, beauty, or creativity, indeed, without a personal existence. Five years later the most popular actor of the period, the great Charlie Chaplin (1889–   ), drove home the same idea in his movie *Modern Times*, which depicted the common man as the hapless and helpless victim of a dehumanized technology. These two artists have set the tone: only by renouncing modern civilization altogether can man survive as man. This theme has since been struck with increasing frequency, and certainly with increasing loudness. The pessimists of today, however, suffer from a bad case of romantic delusion; the "happier society of the preindustrial past" they invoke never existed. In the late thirteenth century, Genghis Khan and his Mongols, whose raids covered an area from China to central Europe, killed as many people—and a much larger proportion of a much smaller population—as two twentieth-century world wars and Hitler put together, yet their technology was the bow and arrow.

However much justice there may be in Huxley's and Charlie Chaplin's thesis, it is sterile. The repudiation of technology they advocate is clearly not the answer. The only positive alternative to destruction by technology is to make technology work as our servant. In the final analysis this surely means mastery by man over himself, for if anyone is to blame, it is not the tool but the human maker and user. "It is a poor carpenter who blames his tools," says an old proverb. It was naive of the nineteenth-century optimist to expect paradise from tools and it is equally naive of the twentieth-century pessimists to make the new tools the scapegoat for such old shortcomings as man's blindness, cruelty, immaturity, greed, and sinful pride.

It is also true that better tools demand a better, more highly skilled, and more careful carpenter. As its ultimate impact on man and his society, twentieth-century technology, by its very mastery of nature, may thus have brought man face to face again with his oldest and greatest challenge: himself.

6 | The Once and Future Manager

The professional manager has not one job, but three. The first is to make economic resources economically productive. The manager has an entrepreneurial job, a job of moving resources from yesterday into tomorrow; a job, not of minimizing risk, but of maximizing opportunity. Every manager spends a very large part of his time with problems that are essentially economic, at least in their results. For instance, where are the markets? How can we achieve a little more productivity from these resources? What are the right things to do, and the right things to stop doing? So everybody who is a manager, no matter whether he is a general manager or a specialist, wrestles for part of his day with an economic dimension.

Then there is a managerial or "administrative" job of making human resources productive, of making people work together, bringing to a common task their individual skills and knowledge; a job of making strengths productive and weaknesses irrelevant, which is the purpose of organization.

These notes from talks given at lectures and seminars in England were first published in *Management Today*, May 1969.

Organization is a machine for maximizing human strengths. If you have a man very good at making things, and no good at marketing and finance, who is in business for himself, you know that he is not going to last very long. If you have an organization, even a small one, you can use a good manufacturing man because you can use his strengths, and his weaknesses are not relevant. You have other people who are good at marketing, or at finance, so that you can build a team in which the strengths of individuals count.

Then there is a third function. Whether they like it or not, managers are not private, in the sense that what they do does not matter. They are public. They are visible. They represent. They stand for something in the community. In fact, they are the only leading group in society—not just the business manager, but all the executives of organizations in this developed, highly organized, highly institutionalized society. Managers have a public function. They may discharge it by a great deal of work outside the business within the community, from Royal Commissions down to the local Boy Scouts troop. Or they may discharge it purely within their own business by leadership and example. But they always do discharge it. Nothing anybody who is a manager does is private, in the sense that one can say: "This is my own affair. It does not concern anybody else. What I do is, therefore, of no real interest to anybody." Managers are on the stage, with the spotlight on them.

So the executive job, as it is today, not as it will be tomorrow, is threefold; a job in which we need objectives, and we need tools; a job in which we need character, and we need competence; a job in which we have to decide, "this we are willing to do, and, therefore, we need to learn how to do it well," or "this we are not going to do, we will let someone else do it, it is beyond our ken, beyond our competence." These are the demarcations of the job.

The Conglomerates Will Be the Stranded Giants of the Next Decade

Whether any one man, or any team of men, can manage a great complexity of different businesses, as in the conglomerates, is a doubtful matter. I came into the world of business a long time ago. My first job in the City of London was to liquidate the stranded giants of the 1920's. I was a pretty good international grave robber. I do not want to go out liquidating the stranded giants of the 1960's. I am afraid, however, that the conglomerates will be the stranded giants of the next decade.

Putting it very bluntly, I do not believe that one can manage a business by reports. I am a figures man, and a quantifier, and one of those people to whom figures talk. I also know that reports are abstractions, and that they can only tell us what we have determined to ask. They are high-level abstractions. That is all right if we have the understanding, the meaning, and the perception. One must spend a great deal of time outside, where the results are. Inside a business one only has costs. One looks at markets, at customers, at society, and at knowledge, all of which are outside the business, to see what is really happening. That reports will never tell you.

At the really critical moment, when a business is in trouble—and I have never seen a business that is not in trouble sooner or later—there is a very high premium on understanding a business, and not just on calculating. So, the conglomerates make me very uneasy, because they put far too much trust in reports. Reports are very comforting to me; they tell me a great deal. But they have also misled me often enough to make me realize that, unless I go out and gain understanding, I may be acting on yesterday, even though the information is up to date.

The belief that one can manage, and invest in, many businesses is based on the assumption that, if things go wrong, one can always sell out and walk away, and let somebody else worry. But I think that to manage—which means being responsible for performance and direction—one has to have a certain core of understanding. If one has a shipping line, and a bank, and an insurance company, and a chocolate company, and a petrochemical company, and a textbook business—I am simply listing the businesses of one company I know—when it comes to the critical moment, one cannot really understand. I do not understand that many businesses. It is hard enough to understand one. I do not understand that many markets, or that many temperaments. The people in a publishing house are very different from the people in a department store—and should be. Being a buyer of ladies' underwear is very different from being a buyer of novels, both in temperament and in knowledge. I have, therefore, very grave doubts about the conglomerates.

At the same time, in this modern world of ours, yesterday's demarcations, yesterday's industry classifications, yesterday's technology lines, no longer apply. They are becoming fuzzy; they interact; they cut across. People who buy packaging do not buy tin cans, and they do not buy paper, or glass; they buy just packaging. They do not care what the material is. On the other hand, if you have a glass company, the only thing which comes out of the kilns is glass; no matter how hard you try, you are not going to produce paper from them. Here is a very real problem, which makes yesterday's industry structures increasingly inappropriate.

The conglomerate builders in America also have an understanding which the old-line managers do not have. They are the first people to understand the new capital market. This has been arriving for the past thirty or forty years. It is a market because a large middle class suddenly arose which had enough surplus money. In my youth, it was an axiom in the City that 99 percent of the people would never have enough surplus to do more than buy life insurance and pay back their mortgages, which are necessities. The capital market was less than 1 percent of the population. Today the capital market in the United Kingdom is probably 25 percent of the population. In America, it is closer to 40 percent. Even on the Continent, it is going up to 10 percent or 15 percent. That is a real market, with choices.

The conglomerate people are the first to understand what that market wants and needs, and they package for that market. However, we have all learned that the first response to a new situation is the wrong response—the right question, but the wrong answer. So I think that the conglomerates are giving the wrong answer, and that we are all going to pay dearly for it, at least in the United States.

I see a need to find a way in which you can adapt yourself to the growing complexity of technology and market, while at the same time maintaining a core of unity that can lie either in the market, or in technology. Here are two examples of the right kinds of conglomerate. Sears, Roebuck is probably the largest retailer in the world. It is willing to buy anything which the American family needs, whether it be fabric, underwear, life insurance, or garden furniture. As long as the family buys it, it is Sears, Roebuck's business, because Sears, Roebuck understands what the family is as an economic unit, and is the expert buyer for the family. The role of the merchant is never to be a seller; it is always to be a buyer for the customer. That is a conglomerate. But, however many different items it carries, it is a unified business.

At the other extreme, Corning Glass is willing to go into any market, as long as it is based on glass technology. It is in the customer market, it is the largest producer of television tubes—any market, as long as it is glass—because they understand their technology. These two extremes both are manageable and make sense. But I am afraid that my friend who is trying to balance the economic risks of a shipping line by having a perfume company is going to be in trouble. In fact, he already is.

Never Look at Any One Measure Alone in Any Business; Look at Multiple Measures

I will never accept anything as "the" right measure of efficiency. Perhaps this is an admission of defeat. I have given up even looking for the right measure. I want multiple measures. When it comes to capital appropriations, I want to see return on capital, and pay-out, and discounted cash flow—all three. Today, this is one of the things I demand of the computer. Ten years ago, this meant that 25,000 clerks with 25,000 pens would have to work 25,000 years to get it. I look at the three and ask what they really tell me. If the Home Office pathologist cuts a hair lengthwise and cross-wise and on the bias, he looks at all three of them under the microscope, so that he will see the one that tells him something about the murder.

I would never start out with earnings per share, because leverage is a very dangerous thing. First, it works both ways, as some of us with older memories will remember, and second, the underlying assumption that a business can be unprofitable or unproductive but my investment in it can be productive, is a very short-range assumption. It is all right if you can sell out after six weeks but not if you are stuck with it. I look at return on total assets as one of the key figures, as I look at return per dollar employed, in other words, the productivity of capital, the value added. But I also look at earnings per share, because after I understand the economics of a business, I then ask how I finance it.

I speak like an old banker, which is exactly what I am, but

you would be surprised how backward the art of finance is and how few industrialists realize how one structures finance. Many businesses use equity capital to finance the production of commodities, which is madness. Commodities are something that the banker will loan money on. Few people realize that once you understand the total economics, then you form the financial structure, using the various money streams, which change all the time. Very often you see businesses where the economics are sound but the financing is wrong, so that the earnings per share are way too low: where one can refinance a business, restructure it, give a business the capacity to attract capital. Sometimes you see the opposite, where everybody in the market is rushing in and buying shares in a business wildly because the earnings per share seem to be going up; and actually it is a low-profit business, very cunningly camouflaged, through financial manipulation. This can last usually for a maximum of eighteen months. Then the stock market suddenly discovers that something is wrong; but, for eighteen months, a lot of people can be fooled. I would never look at one measure alone, on anything in business. First, these measurements are not good enough; second, we do not understand enough for any single measurement to be "the" ultimate measurement.

The First Yardstick by Which Management Is Judged Is, Do They Keep Us Busy?

In every organization you know, there are many people who are being promoted up to the point where they no longer perform. Up to that point, they did well, so they were promoted. When they no longer perform, they are not promoted, but they stay there, we all know this. If it is inevitable—and it *is* inevitable—that we are promoting many people on the basis of performance, up to the point where we promote them beyond their capacity, perhaps this is something which we ought to tackle, instead of just being reconciled to it. The best managers I know spend a good deal of time upon something on which the rest of us spend no time, namely, on thinking through their organizational dilemmas.

Take, for example, the man who started out when the company was small. He was a very good bookkeeper. The company grew, and geological forces raised him to the point where he is now financial V.P. of a very large business—and he is still a bookkeeper. Everybody knows some of these examples, not only in the financial area but in every area. He has been with the company twenty-eight years. He is approaching fifty-five, and he has come in every morning at nine, and has been the last man to leave. Nobody has ever criticized him, and now, suddenly, he is beyond his competence, out of his depth, and a danger to the organization.

What do we do now? Most of us say, "We cannot do anything, so let us try to build around him." The really good managers whom I know do not accept that. They say, "Yes, we owe loyalty where loyalty has been given. We should have taken corrective action long ago, but it is too late now. We should not have let him go up to that position, but it is too late. But we cannot allow him to remain there, because he is doing a great deal of damage." The damage is caused, not because he is not a good financial officer and you need one, but because he tells your organization, "This is what management really expects." He makes cynics out of the young people, and this is one of the sins for which there is no forgiveness.

You cannot fire this man, not because the organization would take a dim view of it, but because most of us are reasonably decent human beings. On the other hand, if you leave him there, you corrupt. So what do you do? Sometimes, one cannot do anything, but say, "All right, we shall have to sweat out the next ten years until he retires." But more often than not, if you really do spend time, you do find a solution that is dignified and considerate. These few cases —they are never very many—are the test of management. It is by this that your organization, your professional, administrative, and management people, right down to the shop floor, really measure you.

An organization measures its management by two yardsticks. The first is "Do they keep us busy? Do they know how to keep us working?" Because if you do not, then you obviously do not take your organization or your own job seriously. The one thing people demand of management is competence. The organization where people are allowed to sit around and mark time has contempt for its management. The other yardstick is "Do they treat the exceptional cases with imagination, intelligence, and compassion?" These are your test cases. Everybody has this proven level of incompetence in their management group. If the man has been with you only five years, you fire him; that is easy. But if he has been with you thirty years, can you move him out where he at least will not do damage? What can you do that is dignified and considerate, and yet tells everybody down the line, "They had his number, and acted on it"?

In the largest organization I know, there are not more than a dozen cases every two or three years. So this is not a large problem in numbers, but it is a big problem in impact. There is no one solution. These cases have to be handled strictly individually. They are the human problems that keep good managers awake at night. By your compassion, but also by your realism, in solving them, your organization will judge you. This is leadership in a business.

The Facts and the Myth of Job Mobility in America Are Not Necessarily the Same

We have plenty of companies in America that advertise for a chemical engineer under forty, with at least forty years' experience; that is very common. The facts, and the myth, of American job mobility are not necessarily quite the same. When you actually analyze one of our large companies, there is a very large amount of turnover and mobility in the management and professional ranks during the first five or seven years. There is a considerable amount at the top, and none in between—none. There we have almost lifetime employment, like the Japanese.

If you actually break it down into market segments, you will find that the young people hop around a great deal. In many cases they have no choice. A great many companies have magnificent personnel policies on paper, and that is all they have. Take a young man who starts out in design engineering. After three years he discovers that this is not what he really wants to do or is good at. His company is advertising for sales engineers, but he has no way of applying, if only because his boss would take a very dim view of any intimation that he might wish to move. So he quits, and the company has only itself to blame.

My students are men of thirty to thirty-three, with six, seven, or ten years' experience, and they come to me and tell me such things. I look at them and say, "Whom do you work for?" They say, "The ABC Company." I say to them, "Across the street from where you work is the employment agency your company uses. You quit, go there, and the next day you will have the job you have been trying to be transferred to for eighteen months." It works every time.

That is not the only reason why the young people are mobile. To be mobile is one way of finding out where you belong. This is not to say that some of them do not overdo it. Then they settle down, they marry, and children come along. The forces that keep them static increase their pull. When they reach top management they may start moving around again.

We have another, smaller problem in another area: the good, technical, functional man of forty-four or forty-five, who has now been Director of Market Research for fifteen years. By now he knows all about the toy market that he is ever going to know, and he is bored. He knows perfectly well that he is never going to be vice-president (marketing); he may want to move, and he should move. Where he is, he is becoming a barnacle, and slowing down the ship to a considerable extent. These are usually timid people, with much at stake in pensions and so forth. There we should have more mobility, in that middle group of purely functional people, who will not rise to general management, and do not wish to do so. They are bored with what they have been doing for too long. They have lost enthusiasm. They have lost any willingness to learn. They know only the right way, the wrong way, and the company way.

Small Business Has Done Much Better Than Any Other in the Last Twenty Years

I have heard for some forty years that the small business is in trouble, and I used to believe it. After twenty years, I said, "Where is the evidence?" I have not seen any. In fact, the small business has done much better than any other in the last twenty years, in every place, including Britain. More small businesses have been started and more small businesses have been prosperous. What is "small" may have changed. But the distribution of businesses has changed amazingly little in the last fifty years, in any major country.

The merger move on today is not really threatening the small business.

Most small businesses believe they need management less; they need management more. A large business can hire a lot of specialists; a small business cannot and, therefore, has to be better at what it is doing. Second, they need objectives much more than do large businesses. They need a realization of what they are really trying to do. They need much more concentration, as they have fewer energies. And they have a different but very serious problem of management succession, precisely because they are usually family companies and because they cannot offer a great deal to the professional nonfamily man, unless they make him an owner, which is not easy with our tax laws. So they have to insist much more rigorously on performance in top jobs. The secret of a family company is a very simple one; as long as you demand that the family members at the top work twice as hard as anybody else, you are all right; but the moment you allow the playboy in management, you are gone, because then the people you employ will not work for you any more, if they are any good. In a family company, subordinates will be perfectly willing to work for a not terribly bright family member, as long as he works.

The real problem in small businesses is not that of being a small business, it is that of the business that outgrows small size; that is where you have your mortality; those are the businesses that are being bought up—the business that has outgrown what the original founder can manage, that by any objective analysis should grow, but bumps against an invisible ceiling all the time. There you have a problem of how to make it possible for a man to change his basic habits, because he strangles the business. Some of them, bluntly, do not want the business to grow.

I have seen businesses where the founding management suddenly realizes that it has three or four or five hundred employees and six markets and now has to build itself a management team, get some information, and think through its own role. The founder realizes that he will have to stop playing at every position and will have to build and encourage and lead players. That is a real crisis of small businesses. It is very difficult for them to grow into medium-sized ones because this is not a matter of quantity, but a matter of basic change in habit, in behavior, and in values.

The Main Impact of the Computer Has Been to Create Unlimited Jobs for Clerks

The computer came on the scene in the late 1940's and, despite all the talk about how fast things go today, we have not yet got an information industry. What we need is not going to be a physical object. It is going to be what is called software—the concepts, the ideas, the logic. There also has to be a lot of peripheral and transmitting and receiving and sending equipment that will make the computer a tool one can use, which it is not today. So far the main impact of the computer has been the creation of unlimited employment opportunities for clerks. This is not great progress. But we are coming very close to the point where we will have an information industry. The pieces are probably all there: the communications satellite and the television screen and the duplicating machine and the fast printer.

What we lack primarily are large concepts which will enable people to use the machine. It will not really become usable as long as we make the asinine attempt to have the computer speak English, which it cannot do. In music, the difference between East and West is the fact that many centuries ago St. Ambrose invented notation. Up to that time music was described in words, as it still is in the East, which means you cannot have ensemble music, you cannot have keys, and you have to memorize. But we all expect seven-year-olds to learn notation in two weeks, and most of them can do it.

We are beginning to learn notation which will essentially enable anybody to use the computer without that unspeakably clumsy, slow, and expensive "programming" or translation job. The proper notation, which will enable us to use an electronic medium electronically, rather than trying to use bastard language that it cannot handle and we cannot handle, is perhaps ten years away.

The future manager will find the computer as much a fact of life as children today find the telephone. This is a new form of energy: information is energy for the mind. What should the manager try to do with it? The first question is, Does it free you? Does it enable you to spend less and less time controlling and more and more time doing the important things? If the result of the computer is that you pore over more records, you are abusing it or you are being abused by it. Then you have less control, incidentally: control is not an abundance of facts, but knowing what facts to have and what they mean.

If it enables you to spend no time controlling operations, because you have thought through what you expect—and, if what you expect does not happen, you know immediately, but, so long as it does happen, you do not have to worry very much—then you are using the computer properly. The first test is, How many hours outside the office does the computer give you? In the office, you are cost-centered and not result-centered. The computer is a tool of liberation if used correctly. Otherwise, you become its servant. It should liberate you from being chained to operations and to your desk and enable you to have time for people and for the outside, where the results are.

The second test is, Are you using it to enable the people in your organization to do what they are ostensibly being paid for? Or are you using it to make it even easier for them to do everything *except* what they are being paid for? There used to be very little choice, but there is no reason for this any more—if you instruct your systems and computer people properly, instead of having them tell you what they should focus on, which is invariably the payroll. Nobody had any difficulty in getting out the payroll before the computer; so use it for the payroll, but do not believe that it is very much of an advance to do the unnecessary three times as fast.

The area in which management in most businesses is today most impeded in getting performance by information and data processing is the field sales force. Sales managers are now so snowed under with all kinds of paper that they do not know who the customers are, they do not train the sales force. They have never been out. Good salesmen are very poor paper handlers: there is almost an inverse relationship between the ability of a salesman to sell and his handwriting legibility. In that room where you have half a dozen or a dozen girls processing the orders that come in from the field, invariably they are all slaving and sweating over the orders from the good salesman: if only because the others do not send in orders. Good salesmen are not paper pushers, and vice versa.

Therefore, this is the area that needs to be looked at. The only resource of a salesman is time. If you find, as you will find, that 70 percent or 80 percent of your salesmen's time is occupied by sending in information that then has to be gone over again, that is one area in which to put the computer to work. The computer people will say that this is not really technically demanding. They are wrong, and you must say, "Never mind, my boy. If you want to do technically demanding jobs, go back to the university. You are on my payroll." Maybe you can say it a little more nicely. I long ago learned not to be nice, because people do not hear it when you hint.

Ask "What are the areas where handling data has become

an end in itself and has been allowed to overgrow the job?"
That is where the data processing people had better go to
work. Then ask "What are the repetitive crises that really
sidetrack the whole organization again and again?" Is it the
annual inventory battle, of which I take a dim view? Or
other repetitive crises that really should not happen—things
that we have not really thought through, that we have not
anticipated? At least now we can build early warning into
the system.

So these are the instructions I give to my computer people. I say, "By now, children, you have learned how to do
payroll; you may even have learned how to do credit; you
may even have learned how to follow an order through the
plant so that one can coordinate plant scheduling with shipping and customer promises." (Although that is something
everybody says he has done, I have yet to see anybody who
really has.) "Fine. You have learned how to do large-scale
clerical work. Now I want you to start working on information."

The Job Which Most Managers Were Brought Up to Spend Most Time on Will Disappear

The literature of management has been concerned, over
the last fifty years, with the management part of the job,
because it was the new thing. This is not going to become
less important, but it is going to become relatively less
urgent. The job which most managers have been brought up
to spend most of their time on will disappear. They will no
longer spend most of their time scratching for a little du-
bious information about what happened yesterday. Accept
the fact that, the day after tomorrow, they will be able to
get it. Our great-great-grandfathers, who started industries,
spent most of their time trying to get a little power. We now
turn a switch. Nobody worries a great deal about where to

get power from. Tomorrow we will no longer have to worry about where we get the other form of energy: the input of the mind, the information input. That will be easy, too.

Now, however, we have to learn a great deal about the entrepreneurial part of the job, to which we have really paid very little attention during these last fifty or sixty years. It is going to be different and quite demanding for two reasons. First, I think it is likely that the last third of this century will be as innovative an era as was the corresponding period of the nineteenth century. We are already getting industries that are based on the knowledge of this century and are quite different; and there is going to be need for a lot of innovation, not just technological, but social and economic as well. At the same time, the pattern of the late nineteenth century, in which you had the individual inventor, who then somehow teamed up with the money man, is unlikely to be repeated.

A great deal of the innovating activity will have to be carried on in existing businesses, where it has not been done so far. By and large, the old folklore which says that existing businesses are incapable of doing the really new things has so far been proved. Even though they all spend a lot of money on research and development, there is not very much to show for it, except some very beautiful buildings in parklike surroundings.

We have to learn to do the job, simply because the economic realities force us to do so. Not only is every single taxation system, in every single developed country, forcing the capital to stay in existing businesses; but also the human resources are there, and it is of the essence of the new industries that the development stage is where you really need men and money. It is not true that inventions become marketable products faster these days. They become marketable products much less fast. In the nineteenth century, within a few months of invention of the electric light bulb and the telephone, on the other side of the Atlantic, you had commercial installation of both in London. This speed we do not have today. This would take ten years of development work today; and the development phase has become far more expensive, and you need far more knowledge for it. Our complexities are greater and this, too, means that existing businesses will very largely have to do the work.

This puts a very great premium on learning systematically to innovate, as part of the management job in the existing business. This is something where all of us really start out on pretty much the same level of nonperformance, so that everybody has a chance of doing it. The technology gap is a thing of the past, simply because, when it comes to the new industries, there are no advantages on one side or the other: it depends on who is going to do the better job of learning how one does this particular kind of work, which is very largely marketing, very largely development. But it also depends on the ability to have two different kinds of organization under the same corporate form, the managerial and the entrepreneurial, which are not organized in the same way. They require that, in our minds, we keep them not separate but, at least, distinct.

If you also want to know in which way your industry is going, what are tomorrow's products and tomorrow's needs, do not look to your home market. It is very unreliable whether it is as big as the United States or as small as Luxembourg. Look to the international market: it is almost totally reliable. It is not true that the United States sets the fashions: this is a fashionable newspaper myth. The world market has been setting the fashions. The real market research today is world market research; we have to learn to see a world market, instead of just national economies.

The idea of the sovereign state as the one central institution—the idea of Hobbes, Locke, and Rousseau—no longer corresponds to reality; all the large, organized, managed, special-purpose institutions of society are autonomous. They can be conducted, they can be led, they can be controlled, to a certain extent; but they cannot be made undone. They are necessities. They are the only way of getting the job done. You can nationalize them, but that does not mean that you control them. On the contrary we have all learned that the one way not to have control is to nationalize something. It is one of the few well-documented experiences of our generation.

Although business is not really that good, and knows it, business is still way ahead of the other institutions, largely because it has been working on the problems longer. So we are coming to be looked upon as a model. Management is a central function, not in business, but in our society, on the performance of which the very existence of the society depends. Therefore, managers, and business managers in particular, suddenly have a dimension added: of exemplar, of leadership. These, then, are the new challenges, the new jobs. How do we make organizations capable of innovating? How do we make knowledge productive? How do we make our business and our industries capable of operating in a very complicated and very dangerous world economy? And what do we really have to do, so that we embody this leadership function, this representative function, this spotlight role of being the most visible, the most articulate, and the most advanced example of this new species, the people who make organization productive for society and individual alike?

Is the Traditional Organization Structure Going to Work Tomorrow as It Has till Now?

There is sufficient reason to wonder whether the traditional organization structure, with which we are all familiar, is going to work tomorrow the way it has worked for the past forty years. Everybody is familiar with the pyramid. We took our organization structure from the military, and so it is a rank-focused structure. When you look at the high-technology and high-knowledge businesses, this structure does not work. You do need the authority of decisions. There has to be somebody who finally can say "yes" or "no," after which the matter rests and debate ceases. You do need an orderly process for on-going work. But ideas do not observe these channels, or they die.

What we see emerging are, essentially, very complex structures, the analogy to which is not mechanical, as it has been in the traditional organization, but biological. There is no biological organization that has only one axis. Biological organizations have at least two, and usually three. Muscles, nerves, the circulatory system—these are all organizing principles. They coexist in very complex relationships. Probably the kindest way to describe what we are doing is to say that we are "fooling around" with systems which maintain an ordered structure, and yet enable a great deal of positioning according to the logic of the job, on the one hand, and the logic of knowledge on the other.

The high-technology companies are simply showing the way. Their problem is very acute. You may have a physicist, next to a cell biologist, next to a communications engineer, and you cannot say that one is more important than the others. In one task, one man is more important; in the next task, another is. So you need to be able to have spontaneous teams, with a high degree of purpose and order and self-discipline, within a framework of orderly decision making and procedure. Though there are examples where this actually works, they are not yet sufficient to enable us to distill the principle. But we can say that it can be done, and is going to be done. As we move from an organization where there were a few people at the top who had all the decision-making power and all the knowledge, while the rest were at their machines, to an organization where the bulk of the people are paid for knowledge input and, above all, for innovation input, we are going to see more of this development.

Free-form organizations, or whatever fancy word you want to use for them, need exceedingly clear objectives—much clearer objectives than the hierarchical, pyramidal organization needs, where the fellow at the top can change his mind and you get, at least on paper, fairly rapid changes all the way down. (You do not, in reality.) Free-form organizations also need a willingness to commit themselves to objectives and to rather demanding performance goals. Otherwise, they degenerate into a debating society.

Second, they require that the people in the group take responsibility for their contribution; they require that the people at the top say, "Look, we are going to leave you alone as much as we can, but one can only delegate what one understands; one cannot delegate what one does not understand. Therefore, if you want autonomy—and we want you to have autonomy—it is your job to think through and tell us what contribution we should hold you accountable for, what are your priorities. Maybe we are going to look at them and say they seem very fine but they make no sense to us, or we are going to look at something and say it is very fine, but we are still responsible for this company and this is not what we are trying to do. But it is your responsibility to take the initiative and to think it through and to focus yourself on the results of the total organization. Maybe you will say that what you really want to work on will not have results until the year 1992. Fine. There are certain things that have that long a lead-time; there is nothing we can do about it; but at least let them be part of our objective and of our goals." Unless you enforce self-discipline, a good time is had by all—but that is all.

Managers Have to Accept That Industrial Relations Will
Become Increasingly Bitter

While the headlines are going to be focused for a long time on the Industrial Relations aspect of the management of people, this is yesterday's rear-guard action and, like all rear-guard actions, it cannot be won. The purpose of a rear-guard action is to enable the main force to get away. Increasingly, the real job will be the mobilization of knowledge and of the knowledge worker. The cost of the people who are being paid to put knowledge to work is very high: not only because they are paid well, but also because they are not usually people who can be used with great versatility. Knowledge is always specialized, always specific. These are also people who either perform very well or do not perform at all. Mediocre knowledge work, as a rule, is not worth having.

But so far most of us still act as if we believe that we can substitute three mediocre clerks for one first-rate knowledge worker. Not only do three mediocre clerks not produce as much as one knowledge worker; three mediocre people produce nothing at all—they only get in each other's way. We are grossly overstaffed and grossly undermanned in most places. Knowledge, in the last analysis, is the only resource of the developed countries. When it comes to willing backs, the underdeveloped countries are way ahead. One cannot compete with the productivity of labor in underdeveloped countries, if they learn a little management.

While we will have to worry about Industrial Relations, therefore, this is, increasingly, going to be a purely negative, a purely defensive area, in which all one can do is hope that one does not lose ground. The opportunity lies in making knowledge productive and thereby making the labor force of yesterday essentially irrelevant and immaterial. This, however, also implies that industrial relations are going to become increasingly bitter. Accept the fact—accept that the industrial worker in the developed world knows that he is dispensable, and his union leader knows it very well. That makes him increasingly bitter and increasingly resistant. The industrial worker, the main beneficiary of the last seventy years of industrial development, suddenly sees his status and his function in the industrial society threatened. We converted the casual laborer of yesterday, who had neither income nor job security, into the machine operator of today, who has both. And those he is going to keep, but not the status and function, the power that he has had: when a Labor government starts talking about trade-union legislation, something has happened that is fundamental.

The problem will not be solved by the old, traditional remedy of worker membership on the board. Wherever we have tried this, it has corrupted a few unionists, and that is all it has done. It has not had an impact on the rank and file, and it has not impeded management. It is more a symbolical than a real thing. I would say, "Do not involve workers in management process. What are the decisions they should take responsibility for that managers are doing and that are only remotely connected with the things for which managers are responsible?"

Almost thirty years ago, when I helped run a liberal arts college, we called in the students and told them that there was a war on, that we were shorthanded and that they were going to have to run certain things: which was practically everything except the teaching and the hiring of faculty and the determination of the curriculum, which we did. But they ran everything else, including the feeding. They made a botch of it the first year—no worse, let me say, than the faculty committee had done. But in the second year they did a good job; there was no problem, and the leaders emerged. They tried a few crazy things and some of them worked, and some of them did not; but they did a fairly responsible job or they went hungry and, after they had gone without meals twice in a row, the feeding arrangements worked. You would be surprised how salutary it was for them to find out that, if you do not plan meals, you do not get any.

How many of the things managers do are only incidental to their job, including a lot of what is plant discipline—shift assignments and so on—that could be left to employees themselves? No doubt, a lot of management people are working on these matters. All right then, have some redundancies.

7 | The First Technological
Revolution and Its Lessons

Aware that we are living in the midst of a technological revolution, we are becoming increasingly concerned with its meaning for the individual and its impact on freedom, on society, and on our political institutions. Side by side with messianic promises of utopia to be ushered in by technology, there are the most dire warnings of man's enslavement by technology, his alienation from himself and from society, and the destruction of all human and political values.

Tremendous though today's technological explosion is, it is hardly greater than the first great revolution technology wrought in human life seven thousand years ago when the first great civilization of man, the irrigation civilization, established itself. First in Mesopotamia, and then in Egypt and in the Indus Valley, and finally in China, there appeared a new society and a new polity: the irrigation city, which then rapidly became the irrigation empire. No other change in man's way of life and in his making a living, not even the changes under way today, so completely revolutionized human society and community. In fact, the irrigation civiliza-

Presidential address to the Society for the History of Technology, December 29, 1965; first published in *Technology and Culture*, Spring 1966.

tions were the beginning of history, if only because they brought writing.

The age of the irrigation civilization was pre-eminently an age of technological innovation. Not until a historical yesterday, the eighteenth century, did technological innovations emerge which were comparable in their scope and impact to those early changes in technology, tools, and processes. Indeed, the technology of man remained essentially unchanged until the eighteenth century insofar as its impact on human life and human society is concerned.

But the irrigation civilizations were not only one of the great ages of technology. They represent also mankind's greatest and most productive age of social and political innovation. The historian of ideas is prone to go back to ancient Greece, to the Old Testament prophets, or to the China of the early dynasties for the sources of the beliefs that still move men to action. But our fundamental social and political institutions antedate political philosophy by several thousand years. They all were conceived and established in the early dawn of the irrigation civilizations. Any one interested in social and governmental institutions and in social and political processes will increasingly have to go back to those early irrigation cities. And, thanks to the work of archeologists and linguists during the last fifty years, we increasingly have the information, we increasingly know what the irrigation civilizations looked like, we increasingly can go back to them for our understanding both of antiquity and of modern society. For essentially our present-day social and political institutions, practically without exception, were then created and established. Here are a few examples.

(1) The irrigation city first established government as a distinct and permanent institution. It established an impersonal government with a clear hierarchical structure in which very soon there arose a genuine bureaucracy—which

is, of course, what enabled the irrigation cities to become irrigation empires.

Even more basic: the irrigation city first conceived of man as a citizen. It had to go beyond the narrow bounds of tribe and clan and had to weld people of very different origins and blood into one community. This required the first supra-tribal deity, the god of the city. It also required the first clear distinction between custom and law and the development of an impersonal, abstract, codified legal system. Indeed, practically all legal concepts, whether of criminal or of civil law, go back to the irrigation city. The first great code of law, that of Hammurabi, almost four thousand years ago, would still be applicable to a good deal of legal business in today's highly developed, industrial society.

The irrigation city also first developed a standing army—it had to. For the farmer was defenseless and vulnerable and, above all, immobile. The irrigation city which, thanks to its technology, produced a surplus, for the first time in human affairs, was a most attractive target for the barbarian outside the gates, the tribal nomads of steppe and desert. And with the army came specific fighting technology and fighting equipment: the war horse and the chariot, the lance and the shield, armor and the catapult.

(2) It was in the irrigation city that social classes first developed. It needed people permanently engaged in producing the farm products on which all the city lived; it needed farmers. It needed soldiers to defend them. And it needed a governing class with knowledge, that is, originally a priestly class. Down to the end of the nineteenth century these three "estates" were still considered basic in society.*

But at the same time the irrigation city went in for spe-

---

* See the brilliant though one-sided book by Karl A. Wittvogel, *Oriental Despotism: A Comparative Study of Total Power* (New Haven, Conn., 1957).

cialization of labor resulting in the emergence of artisans and craftsmen: potters, weavers, metalworkers, and so on; and of professional people: scribes, lawyers, judges, physicians.

And because it produced a surplus, it first engaged in organized trade which brought with it not only the merchant but money, credit, and a law that extended beyond the city to give protection, predictability, and justice to the stranger, the trader from far away. This, by the way, also made necessary international relations and international law. In fact, there is not very much difference between a nineteenth-century trade treaty and the trade treaties of the irrigation empires of antiquity.

(3) The irrigation city first had knowledge, organized it, and institutionalized it. Both because it required considerable knowledge to construct and maintain the complex engineering works that regulated the vital water supply and because it had to manage complex economic transactions stretching over many years and over hundreds of miles, the irrigation city needed records, and this, of course, meant writing. It needed astronomical data, as it depended on a calendar. It needed means of navigating across sea or desert. It, therefore, had to organize both the supply of the needed information and its processing into learnable and teachable knowledge. As a result, the irrigation city developed the first schools and the first teachers. It developed the first systematic observation of natural phenomena, indeed, the first approach to nature as something outside of and different from man and governed by its own rational and independent laws.

(4) Finally, the irrigation city created the individual. Outside the city, as we can still see from those tribal communities that have survived to our days, only the tribe had existence. The individual as such was neither seen nor paid attention to. In the irrigation city of antiquity, however, the

individual became, of necessity, the focal point. And with this emerged not only compassion and the concept of justice; with it emerged the arts as we know them, the poets, and eventually the world religions and the philosophers.

This is, of course, not even the barest sketch. All I wanted to suggest is the scope and magnitude of social and political innovation that underlay the rise of the irrigation civilizations. All I wanted to stress is that the irrigation city was essentially "modern," as we have understood the term, and that, until today, history largely consisted in building on the foundations laid five thousand or more years ago. In fact, one can argue that human history, in the last five thousand years, has largely been an extension of the social and political institutions of the irrigation city to larger and larger areas, that is, to all areas on the globe where water supply is adequate for the systematic tilling of the soil. In its beginnings, the irrigation city was the oasis in a tribal, nomadic world. By 1900 it was the tribal, nomadic world that had become the exception.

The irrigation civilization was based squarely upon a technological revolution. It can with justice be called a "technological polity." All its institutions were responses to opportunities and challenges that new technology offered. All its institutions were essentially aimed at making the new technology most productive.

I hope you will allow me one diversion.

The history of the irrigation civilizations has yet to be written. There is a tremendous amount of material available now, where fifty years ago we had, at best, fragments. There are splendid discussions available of this or that irrigation civilization, for instance of Sumer. But the very big job of re-creating this great achievement of man and of telling the story of his first great civilization is yet ahead of us.

This should be pre-eminently a job for historians of tech-

nology such as we profess to be. At the very least the job
calls for a historian with high interest in, and genuine understanding of, technology. The essential theme around which
this history will have to be written must be the impacts and
capacities of the new technology and the opportunities and
challenges which this, the first great technological revolution, presented. The social, political, cultural institutions,
familiar though they are to us today—for they are in large
measure the institutions we have been living with for five
thousand years—were all brand-new then, and were all the
outgrowth of new technology and of attempts to solve the
problems the new technology posed.

It is our contention in the Society for the History of Technology that the history of technology is a major, distinct
strand in the web of human history. We believe that the
history of mankind cannot be properly understood without
relating to it the history of man's work and man's tools, that
is, the history of technology. Some of our colleagues and
friends—let me mention only such familiar names as Lewis
Mumford, Fairfield Osborn, Joseph Needham, R. J. Forbes,
Cyril Stanley Smith, and Lynn White—have in their own
works brilliantly demonstrated the profound impact of technology on political, social, economic, and cultural history.
But while technological change has always had impact on
the way men live and work, surely at no other time has
technology so literally shaped civilization and culture as
during the first technological revolution, that is, during the
rise of the irrigation civilizations of antiquity.

Only now, however, is it possible to tell the story. No
longer can its neglect be justified. For the facts are available,
as I stated before. And we now, because we live in a technological revolution ourselves, are capable of understanding
what happened then—at the very dawn of history. There is a
big job to be done: to show that the traditional approach to
our history—the approach taught in our schools—in which

"relevant" history really begins with the Greeks (or with the Chinese dynasties), is shortsighted and distorts the real "ancient civilization."

I have, however, strayed off my topic: the question I posed at the beginning, what we can learn from the first technological revolution regarding the impacts likely to result on man, his society, and his government from the new industrial revolution, the one we are living in. Does the story of the irrigation civilization show man to be determined by his technical achievements, in thrall to them, coerced by them? Or does it show him capable of using technology to his own, to human ends, and of being the master of the tools of his own devising?

The answer which the irrigation civilizations give us to this question is threefold.

(1) Without a shadow of doubt, major technological change creates the need for social and political innovation. It does make obsolete existing institutional arrangements. It does require new and very different institutions of community, society, and government. To this extent there can be no doubt: technological change of a revolutionary character coerces; it *demands innovation.*

(2) The second answer also implies a strong necessity. There is little doubt, one would conclude from looking at the irrigation civilizations, that specific technological changes demand equally specific social and political innovations. That the basic institutions of the irrigation cities of the Old World, despite great cultural difference, all exhibited striking similarity may not prove much. After all, there probably was a great deal of cultural diffusion (though I refuse to get into the quicksand of debating whether Mesopotamia or China was the original innovator). But the fact that the irrigation civilizations of the New World around the Lake of Mexico and in Maya Yucatan, though culturally completely

independent, millennia later evolved institutions which, in fundamentals, closely resemble those of the Old World (e.g., an organized government with social classes and a permanent military, and writing) would argue strongly that the solutions to specific conditions created by new technology have to be specific and are, therefore, limited in number and scope.

In other words, one lesson to be learned from the first technological revolution is that new technology creates what a philosopher of history might call "objective reality." And objective reality has to be dealt with on *its* terms. Such a reality would, for instance, be the conversion, in the course of the first technological revolution, of human space from "habitat" into "settlement," that is, into a permanent territorial unit always to be found in the same place—unlike the migrating herds of pastoral people or the hunting grounds of primitive tribes. This alone made obsolete the tribe and demanded a permanent, impersonal, and rather powerful government.

(3) But the irrigation civilizations can teach us also that the new objective reality determines only the gross parameters of the solutions. It determines where, and in respect to what, new institutions are needed. It does not make anything "inevitable." It leaves wide open *how* the new problems are to be tackled, what the purposes and values of the new institutions are to be.

In the irrigation civilizations of the New World the individual, for instance, failed to make his appearance. Never, as far as we know, did these civilizations get around to separating law from custom nor, despite a highly developed trade, did they invent money.

Even within the Old World, where one irrigation civilization could learn from the others, there were very great differences. They were far from homogeneous even though all had similar tasks to accomplish and developed similar institutions for these tasks. The different specific answers expressed above all different views regarding man, his position in the universe, and his society—different purposes and greatly differing values.

Impersonal bureaucratic government had to arise in all these civilizations; without it they could not have functioned. But in the Near East it was seen at a very early stage that such a government could serve equally to exploit and hold down the common man and to establish justice for all and protection for the weak. From the beginning the Near East saw an ethical decision as crucial to government. In Egypt, however, this decision was never seen. The question of the purpose of government was never asked. And the central quest of government in China was not justice but harmony.

It was in Egypt that the individual first emerged, as witness the many statues, portraits, and writings of professional men, such as scribes and administrators, that have come down to us—most of them superbly aware of the uniqueness of the individual and clearly asserting his primacy. It is early Egypt, for instance, which records the names of architects who built the great pyramids. We have no names for the equally great architects of the castles and palaces of Assur or Babylon, let alone for the early architects of China. But Egypt suppressed the individual after a fairly short period during which he flowered (perhaps as part of the reaction against the dangerous heresies of Ikhnaton). There is no individual left in the records of the Middle and New Kingdoms, which perhaps explains their relative sterility.

In the other areas two entirely different basic approaches emerged. One, that of Mesopotamia and of the Taoists, we might call "personalism," the approach that found its greatest expression later in the Hebrew prophets and in the Greek dramatists. Here the stress is on developing to the fullest the capacities of the person. In the other approach—we might call it "rationalism," taught and exemplified above all by Confucius—the aim is the moulding and shaping of the individual according to pre-established ideals of rightness and perfection. I need not tell you that both these approaches still permeate our thinking about education.

Or take the military. Organized defense was a necessity for the irrigation civilization. But three different approaches emerged: a separate military class supported through tribute by the producing class, the farmers; the citizen-army drafted from the peasantry itself; and mercenaries. There is very little doubt that from the beginning it was clearly understood that each of these three approaches had very real political consequences. It is hardly coincidence, I believe, that Egypt, originally unified by overthrowing local, petty chieftains, never developed afterward a professional, permanent military class.

Even the class structure, though it characterizes all irrigation civilizations, showed great differences from culture to culture and within the same culture at different times. It was being used to create permanent castes and complete social immobility, but it was also used with great skill to create a very high degree of social mobility and a substantial measure of opportunities for the gifted and ambitious.

Or take science. We now know that no early civilization excelled China in the quality and quantity of scientific observations. And yet we also know that early Chinese culture did not point toward anything we would call science. Perhaps because of their rationalism the Chinese refrained from generalization. And though fanciful and speculative, it is the generalizations of the Near East and the mathematics of Egypt which point the way toward systematic science. The Chinese, with their superb gift for accurate observation, could obtain an enormous amount of information about nature. But their view of the universe remained totally unaffected thereby—in sharp contrast to what we know about the Middle Eastern developments out of which Europe arose.

In brief, the history of man's first technological revolution indicates the following:

(1) Technological revolutions create an objective need for social and political innovations. They create a need also for identifying the areas in which new institutions are needed and old ones are becoming obsolete.

(2) The new institutions have to be appropriate to specific new needs. There are right social and political responses to technology and wrong social and political responses. To the extent that only a right institutional response will do, society and government are largely circumscribed by new technology.

(3) But the values these institutions attempt to realize, the human and social purposes to which they are applied, and, perhaps most important, the emphasis and stress laid on one purpose as against another, are largely within human control. The bony structure, the hard stuff of a society, is prescribed by the tasks it has to accomplish. But the ethos of the society is in man's hands and is largely a matter of the "how" rather than of the "what."

For the first time in thousands of years, we face again a situation that can be compared with what our remote ancestors faced at the time of the irrigation civilization. It is not only the speed of technological change that creates a revolution, it is its scope as well. Above all, today, as seven thousand years ago, technological developments from a great many areas are growing together to create a new human environment. This has not been true of any period between the first technological revolution and the technological revolution that got under way two hundred years ago and has still clearly not run its course.

We, therefore, face a big task of identifying the areas in which social and political innovations are needed. We face a big task in developing the institutions for the new tasks, institutions adequate to the new needs and to the new capacities which technological change is casting up. And, finally, we face the biggest task of them all, the task of insuring that the new institutions embody the values we believe in, aspire to the purposes we consider right, and serve human freedom, human dignity, and human ends.

If an educated man of those days of the first technological revolution—an educated Sumerian perhaps or an educated ancient Chinese—looked at us today, he would certainly be totally stumped by our technology. But he would, I am sure, find our existing social and political institutions reasonably familiar—they are after all, by and large, not fundamentally different from the institutions he and his contemporaries first fashioned. And, I am quite certain, he would have nothing but a wry smile for both those among us who predict a technological heaven and those who predict a technological hell of "alienation," of "technological unemployment," and so on. He might well mutter to himself, "This is where I came in." But to us he might well say, "A time such as was mine and such as is yours, a time of true technological revolution, is not a time for exultation. It is not a time for despair either. It is a time for work and for responsibility."

It is easier to define long-range planning by what it is not rather than by what it is. Three things in particular, which it is commonly believed to be, it emphatically is not.

(1) First, it is not "forecasting." It is not masterminding the future, in other words. Any attempt to do so is foolish; human beings can neither predict nor control the future.

If anyone still suffers from the delusion that the ability to forecast beyond the shortest time span is given to us, let him look at the headlines in yesterday's paper, and then ask himself which of them he could possibly have predicted ten years ago.

Could he have forecast that by today the Russians would have drawn even with us in the most advanced branches of physical sciences and of engineering? Could he have forecast that West Germany, in complete ruins and chaos then, would have become the most conservative country in the world and one of the most productive ones, let alone

This article, reprinted from *Management Science,* vol. 5, no. 3 (April 1959) is based on a paper given before the Fourth International Meeting of the Institute of Management Sciences, held in Detroit, October 17–18, 1957.

that it would become very stable politically? Could he have forecast that the Near East would become a central trouble spot, or would he have had to assume that the oil revenues there would take care of all problems?

This is the way the future always behaves. To try to mastermind it is, therefore, childish; we can only discredit what we are doing by attempting it. We must start out with the conclusion that forecasting is not respectable and not worthwhile beyond the shortest of periods. *Long-range planning is necessary precisely because we cannot forecast.*

But there is another, and even more compelling, reason why forecasting is not long-range planning. Forecasting attempts to find the most probable course of events, or at best, a range of probabilities. But the entrepreneurial problem is the unique event that will change the possibilities, for the entrepreneurial universe is not a physical- but a value-universe. Indeed, the central entrepreneurial contribution, and the one which alone is rewarded with a profit, is to bring about the unique event, the *innovation* that changes the probabilities.

Let me give an example—a very elementary one which has nothing to do with innovation but which illustrates the importance of the improbable even for purely adaptive business-behavior.

A large coffee distributor has for many years struggled with the problem of the location and capacity of its processing plants throughout the country. It had long been known that coffee prices were as important a factor in this, as location of market, volume, or transportation and delivery strategy. Now if we can forecast anything, it is single-commodity prices; and the price forecasts of the company economists have been remarkably accurate. Yet the decisions on plant location and capacity based on these forecasts have again and again proven costly blunders. Extreme pricing events, the probability of which at any one time was exceedingly low, had, even if they lasted only for a week at a time, impact on the economics of the system that were vastly greater than that of the accurately forecast "averages." Forecasting, in other words, obscured economic reality. What was needed (as the Theory of Games could have proven) was to look at the extreme possibilities, and to ask, "Which of these can we not afford to disregard?"

The only thing atypical in this example is that it is so simple. Usually things are quite a bit more complex. But despite its (deceptive) simplicity it shows why forecasting is not an adequate basis even for purely adaptive behavior, let alone for the entrepreneurial decisions of long-range planning.

(2) The next thing to be said about what long-range planning is not, is that it does not deal with future decisions. It deals with the *futurity of present decisions.*

Decisions exist only in the present. The question that faces the long-range planner is not what we should do tomorrow. It is: what do we have to do today to be ready for an uncertain tomorrow? The question is not what will happen in the future. It is: what futurity do we have to factor into our present thinking and doing, what time-spans do we have to consider, and how do we converge them to a simultaneous decision in the present?

Decision making is essentially a time machine which synchronizes into one present a great number of divergent time-spans. This is, I think, something which we are only learning now. Our approach today still tends toward the making of plans for something we will decide to do in the future. This may be a very entertaining exercise, but it is a futile one.

Again, long-range planning is necessary because we can make decisions only *in* the present; the rest are pious intentions. And yet we cannot make decisions *for* the present alone; the most expedient, most opportunist decision—let alone the decision not to decide—may commit us on a long-range basis, if not permanently and irrevocably.

(3) Finally, the most common misconception of all, *long-range planning is not an attempt to eliminate risk*. It is not even an attempt to minimize risk. Indeed, any such attempt can only lead to irrational and unlimited risk and to certain disaster.

The central fact about economic activity is that, by definition, it commits present resources to future and, therefore, highly uncertain expectations. To take risk is, therefore, the essence of economic activity. Indeed, one of the most rigorous theorems of economics (Boehm-Bawerk's Law) proves that existing means of production will yield greater economic performance only through greater uncertainty, that is, through greater risk.

But while it is futile to try to eliminate risk, and questionable to try to minimize it, it is essential that the risks taken be the *right risks*. The end result of successful long-range planning must be a capacity to take a greater risk; for this is the only way to improve *entrepreneurial* performance. To do this, however, we must know and understand the risks we take. We must be able to rationally choose among risk-taking courses of action rather than plunge into uncertainty on the basis of hunch, hearsay, or experience (no matter how meticulously quantified).

Now I think we can attempt to define what long-range planning is. It is the continuous process of making *present entrepreneurial (risk-taking) decisions* systematically and with the best possible knowledge of their futurity, organizing systematically *the efforts* needed to carry out these decisions, and measuring the results of these decisions against the expectations through *organized, systematic feedback*.

"This is all very well," many experienced businessmen might say (and do say). "But why make a production out of it? Isn't this what the entrepreneur has been doing all along, and doing quite successfully? Why, then, should it need all this elaborate mumbo jumbo? Why should it be an organized, perhaps even a separate activity? Why, in other words, should we even talk about 'long-range planning,' let alone do it?"

It is perfectly true that there is nothing very new to entrepreneurial decisions. They have been made as long as we have had entrepreneurs. There is nothing new in here regarding the essentials of economic activity. It has always been the commitment of present resources to future expectations; and for the last three hundred years this has been done in contemplation of change. (This was not true earlier. Earlier economic activity was based on the assumption that there would be no change, which assumption was institutionally guarded and defended. Altogether up to the seventeenth century it was the purpose of all human institutions to prevent change. The business enterprise is a significant and rather amazing novelty in that it is the first human institution having the purpose of bringing about change.)

But there are several things which are new; and they have created the need for the organized, systematic, and, above all, specific process that we call "long-range planning."*

---

* "Long-range planning" is not a term I like or would have picked myself. It is a misnomer—as are so many of our terms in economics and management, such as "capitalism," "automation," "operations research," "industrial engineering," or "depreciation." But it is too late to do anything about the term; it has become common usage.

(1) The time-span of entrepreneurial and managerial decisions has been lengthening so fast and so much as to make necessary systematic exploration of the uncertainty and risk of decisions.

In 1888 or thereabouts, and old and perhaps apocryphal story goes, the great Thomas Edison, already a world figure, went to one of the big banks in New York for a loan on something he was working on. He had plenty of collateral and he was a great man; so the vice-presidents all bowed and said "Certainly, Mr. Edison, how much do you need?" But one of them, out of idle curiosity asked, "Tell me, Mr. Edison, how long will it be before you have this new product?" Edison looked him in the eye and said, "Son, judging from past experience, it will be about eighteen months before I even know whether I'll have a product or not." Whereupon the vice-presidents collapsed in a body, and, despite the collateral, turned down the loan application. The man was obviously mad; eighteen months of uncertainty was surely not a risk a sane businessman would take!

Today practically every manager takes ten or twenty year risks without wincing. He takes them in product development, in research, in market development, in the development of a sales organization, and in almost anything. This lengthening of the time-span of commitment is one of the most significant features of our age. It underlies our economic advances. But while quantitative in itself, it has changed the qualitative character of enterpreneurial decisions. It has, so to speak, converted time from being a dimension in which business decisions are being made into an essential element of the decisions themselves.

(2) Another new feature is the speed and risk of innovation. To define what we mean by this term would go far beyond the scope of this paper.*

But we do not need to know more than that industrial research expenditures (that is, business expenditures aimed at innovating primarily peacetime products and processes) have increased in this country from less than $100 million in 1928 to $7 or 8 billion in 1958. Clearly, a technologically slow-moving, if not essentially stable, economy has become one of violent technological flux, rapid obsolescence, and great uncertainty.

(3) Then there is the growing complexity both of the business enterprise internally, and of the economy and society in which it exists. There is the growing specialization of work which creates increasing need for common vision, common understanding, and common language, without which top-management decisions, however right, will never become effective action.

(4) Finally—a subtle, but perhaps the most important point—the typical businessman's concept of the basis of entrepreneurial decision is, after all, a misconception.

Most businessmen still believe that these decisions are made by "top management." Indeed, practically all textbooks lay down the dictum that "basic policy decisions" are the "prerogative of top management." At most, top management "delegates" certain decisions.

But this reflects yesterday's rather than today's reality, let alone that of tomorrow. It is perfectly true that top management must have the final say, the final responsibility. But the business enterprise of today is no longer an organization in which there are a handful of "bosses" at the top who make

* For discussion, see my book, *The Landmarks of Tomorrow* (Harper and Brothers, New York, 1958).

all the decisions while the "workers" carry out orders. It is primarily an organization* of professionals of highly specialized knowledge exercising autonomous, responsible judgment. And every one of them—whether manager or individual expert contributor—constantly makes truly entrepreneurial decisions, that is, decisions which affect the economic characteristics and risks of the entire enterprise. He makes them not by "delegation from above" but inevitably in the performance of his own job and work.

For this organization to be functioning, two things are needed: knowledge by the entire organization of what the direction, the goals, the expectations are; and knowledge by top management as to what the decisions, commitments, and efforts of the people in the organization are. The needed focus—one might call it a *model of the relevants in internal and external environment*—only a "long-range plan" can provide.

One way to summarize what is new and different in the process of entrepreneurial decision making is in terms of information. The amount, diversity, and ambiguity of the information that is beating in on the decision maker have all been increasing so much that the built-in experience reaction that a good manager has cannot handle it. He breaks down; and his breakdown will take either of the two forms known to any experimental psychologist. One is withdrawal from reality, i.e., "I know what I know and I only go by it; the rest is quite irrelevant and I won't even look at it." Or there is a feeling that the universe has become completely irrational so that one decision is as good as the other, resulting in paralysis. We see both in executives who have to make decisions today. Neither is likely to result in rational or in successful decisions.

* For a discussion of this "new organization," see again my *Landmarks of Tomorrow*.

There is something else managers and management scientists might learn from the psychologists. Organization of information is often more important to the ability to perceive and act than analysis and understanding of the information. I recall one experience with the organization of research-planning in a pharmaceutical company. The attempt to analyze the research decisions—even to define alternatives of decisions—was a dismal failure. In the attempt, however, the decisions were classified to the point where the research people could know what kind of a decision was possible at what stage. They still did not know what factors should or should not be considered in a given decision, nor what its risks were. They could not explain why they made this decision rather than another one, nor spell out what they expected. But the mere organization of this information enabled them again to apply their experience and to "play hunches"—with measurable and very significant improvement in the performance of the entire research group.

"Long-range planning" is more than organization and analysis of information; it is a decision-making process. But even the information job cannot be done except as part of an organized planning effort—otherwise there is no way of determining which information is relevant.

What, then, are the requirements of long-range planning? We cannot satisfy all of them as yet with any degree of competence; but we can specify them.

Indeed, we can—and should—give two sets of specifications: one in terms of the characteristics of the process itself; another in terms of its major and specific new-knowledge content.

(1) Risk-taking entrepreneurial decisions, no matter whether made rationally or by tea-leaf reading, always embody the same eight elements:

(a) *Objectives.* This is, admittedly, an elusive term, perhaps even a metaphysical one. It may be as difficult for Management Science to define "objectives" as it is for biology to define "life." Yet, we will be as unable to do without objectives as the biologists are unable to do without life. Any entrepreneurial decision, let alone the integrated decision-system we call a "long-range plan," has objectives, consciously or not.

(b) *Assumptions.* These are what is believed by the people who make and carry out decisions to be "real" in the internal and external universe of the business.

(c) *Expectations*—the future events or results considered likely or attainable.

These three elements can be said to *define the decision.*

(d) *Alternative courses of action.* There never is—indeed, in a true uncertainty situation there never can be—"one right decision." There cannot even be "one best decision." There are always "wrong decisions," that is, decisions inadequate to the objectives, incompatible with the assumptions, or grossly improbable in the light of the expectations. But once these have been eliminated, there will still be alternatives left—each a different configuration of objectives, assumptions and expectations, each with its own risks and its own ratio between risks and rewards, each with its own impact, its specific efforts, and its own results. Every decision is thus a value-judgment—it is not the "facts that decide"; people have to choose between imperfect alternatives on the basis of uncertain knowledge and fragmentary understanding.

Two alternatives deserve special mention, if only because they have to be considered in almost every case. One is the alternative of no action (which is, of course, what postponing a decision often amounts to); the other is the very important choice between adaptive and innovating action—each having risks that differ greatly in character though not necessarily in magnitude.

(e) The next element in the decision-making process is the *decision itself*.

(f) But there is no such thing as one isolated decision; every decision is, of necessity, part of a *decision-structure*.

Every financial man knows, for instance, that the original capital appropriation on a new investment implies a commitment to future- and usually larger-capital appropriations which, however, are almost never as much as mentioned in the proposal submitted. Few of them seem to realize, however, that this implies not only a positive commitment but also, by mortgaging future capital resources, limits future freedom of action. The structuring impact of a decision is even greater in respect to allocations of scarce manpower, such as research people.

(g) A decision is only pious intention unless it leads to action. Every decision, therefore, has an *impact stage*.

This impact always follows Newton's Second Law, so to speak; it consists of action and reaction. It requires effort. But it also dislocates. There is, therefore, always the question: what effort is required, by whom, and where? What must people know, what must they do, and what must they achieve? But there is also the question—generally neglected —what does this decision do to other areas? Where does it shift the burden, the weaknesses, and the stress points; and what impact does it have on the outside; in the market, in the supply structure, in the community, and so on?

(h) And, finally, there are *results*.

Each of these elements of the process deserves an entire book by itself. But I think I have said enough to show that both, the process itself and each element in it, are *rational*, no matter how irrational and arbitrary they may appear. Both the process and all its elements can, therefore, be defined, can be studied, and can be analyzed. And both can be improved through systematic and organized work. In particular, as in all rational processes, the entire process is improved and strengthened as we define, clarify, and analyze each of its constituent elements.

(2) We can also, as said above, describe long-range planning in terms of its specific new-knowledge content.

Among the areas where such new knowledge is particularly cogent, might be mentioned:

(a) *The time dimensions of planning*

To say "long-range" or "short-range" planning implies that a given time-span defines the planning; and this is actually how businesses look at it when they speak of a "five-year plan" or a "ten-year plan." But the essence of planning is to make present decisions with knowledge of their futurity. It is the futurity that determines the time span, and not vice versa.

Strictly speaking, "short range" and "long range" do not describe time-spans but stages in every decision. "Short-range" is the stage before the decision has become fully effective, the stage during which it is only "costs" and not yet "results." The "short range" of a decision to build a steel mill are the five years or so until the mill is in production. And the "long-range" of any decision is the period of expected performance needed to make the decision a successful one—the twenty or more years of above break-even point operations in the case of the steel mill, for instance.

There are limitations on futurity. In business decisions the most precise mathematical statement is often that of my eighth-grade teacher that parallels are two lines which do not meet this side of the schoolyard. Certainly, in the expectations and anticipations of a business, the old rule of statistics usually applies—that anything beyond twenty years equals infinity; and since expectations more than twenty years hence have normally a present value of zero, they should receive normally only a minimal allocation of present efforts and resources.

Yet it is also true that, if future results require a long gestation period, they will be obtained only if initiated early enough. Hence, long-range planning requires knowledge of futurity: What do we have to do today if we want to be some place in the future? What will not get done at all if we do not commit resources to it today?

If we know that it takes ninety-nine years to grow Douglas firs in the Northwest to pulping size, planting seedlings today is the only way we can provide for pulp supply in ninety-nine years. Some one may well develop some speeding-up hormone; but we cannot bank on it if we are in the paper industry. It is quite conceivable, may, indeed, be highly probable, that we will use trees primarily as a source of chemicals long before these trees grow to maturity. We may even get the bulk of paper supply thirty years hence from less precious, less highly structured sources of cellulose than a tree, which is the most advanced chemical factory in the plant kingdom. This simply means, however, that our forests may put us into the chemical industry some time within the next thirty years; and we had better learn now something about chemistry. If our paper plants depend on Douglas fir, our planning cannot confine itself to twenty years, but must consider ninety-nine years. For we must be able to say whether we have to plant trees today, or whether we can postpone this expensive job.

But on other decisions even five years would be absurdly long. If our business is buying up distress merchandise and selling it at auction, then next week's clearance sale is "long-range future"; and anything beyond is largely irrelevant to us.

It is the nature of the business and the nature of the decision which determine the time-spans of planning.

Yet the time-spans are not static or "given." The time decision itself is the first and a highly important risk-taking decision in the planning process. It largely determines the allocation of resources and efforts. It largely determines the risks taken (and one cannot repeat too often that to postpone a decision is in itself a risk-taking and often irrevocable decision). Indeed, the time decision largely determines the character and nature of the business.

(b) *Decision structure and configuration*

The problem of the time dimension is closely tied in with that of decision structure.

Underlying the whole concept of long-range planning are two simple insights.

We need an integrated decision structure for the business as a whole. There are really no isolated decisions on a product, or on markets, or on people. Each major risk-taking decision has impact throughout the whole; and no decision is isolated in time. Every decision is a move in a chess game, except that the rules of enterprise are by no means as clearly defined. There is no finite "board" and the pieces are neither as neatly distinguished nor as few in number. Every move opens some future opportunities for decision, and forecloses others. Every move, therefore, commits positively and negatively.

Let me illustrate these insights with a simple example, that of a major steel company today.

I posit that it is reasonably clear to any student of technology (not of steel technology but of technology in general) that steelmaking is on the threshold of major technological change. *What* they are perhaps the steelmaker knows, but *that* they are I think any study of the pattern, rhythm, and, I would say, morphology of technological development, might indicate. A logical—rather than metallurgical—analysis of the process would even indicate *where* the changes are likely to occur. At the same time, the steel company faces the need of building new capacity if it wants to keep its share of the market, assuming that steel consumption will continue to increase. A decision to build a plant today, when there is nothing available but the old technology, means in effect that for fifteen to twenty years the company cannot go into the new technology except at prohibitive cost. It is very unlikely, looking at the technological pattern, that these changes will be satisfied by minor modifications in existing facilities; they are likely to require new facilities to a large extent. By building today the company closes certain opportunities to itself, or at least it very greatly raises the future entrance price. At the same time, by making the decision to postpone building, it may foreclose other opportunities such as market position, perhaps irrevocably. Management therefore has to understand—without perhaps too much detail—the location of this decision in the continuing process of entrepreneurial decision.

At the same time, entrepreneurial decisions must be fundamentally expedient decisions. It is not only impossible to know all the contingent effects of a decision, even for the

shortest time period ahead. The very attempt to know
them would lead to complete paralysis.

But the determination of what should be considered and
what should be ignored, is in itself a difficult and conse-
quential decision. We need knowledge to make it—I might
say that we need a theory of entrepreneurial inference.

(c) *The characteristics of risks*

It is not only magnitude of risk that we need to be able to
appraise in entrepreneurial decisions. It is above all the
character of the risk. Is it, for instance, the kind of risk we
can afford to take, or the kind of risk we cannot afford to
take? Or is it that rare but singularly important risk, the risk
we cannot afford not to take—sometimes regardless of the
odds?

> The best General Electric scientists, we are told, ad-
> vised their management in 1945 that it would be at least
> forty years before nuclear energy could be used to pro-
> duce electric power commercially. Yet General Electric—
> rightly—decided that it had to get into the atomic energy
> field. It could not afford not to take the risk as long as
> there was the remotest possibility that atomic energy
> would, after all, become a feasible source of electric
> power.

We know from experience that the risk we cannot afford
not to take is like a "high-low" poker game. A middle hand
will inevitably lose out. But we do not know why this is so.
And the other, and much more common, kinds of risk we do
not really understand at all.

(d) *Finally, there is the area of measurements*

I do not have to explain to readers of *Management Science* why measurements are needed in management, and
especially for the organized entrepreneurial decisions we
call "long-range planning."

But it should be said that in human institutions, such as a business enterprise, measurements, strictly speaking, do not and cannot exist. It is the definition of a measurement that it be impersonal and objective, that is, extraneous to the event measured. A child's growth is not dependent on the yardstick or influenced by being recorded. But any measurement in a business enterprise determines action—both on the part of the measurer and the measured—and thereby directs, limits, and causes behavior and performance of the enterprise. Measurement in the enterprise is always motivation, that is, moral force, as much as it is *ratio cognoscendi.*

In addition, in long-range planning we do not deal with observable events. We deal with future events, that is, with expectations. And expectations, being incapable of being observed, are never "facts" and cannot be measured.

Measurements, in long-range planning, thus present very real problems, especially conceptual ones. Yet precisely because what we measure and how we measure determines what will be considered relevant, and determines thereby not just what we see, but what we—and others—do, measurements are all-important in the planning process. Above all, unless we build expectations into the planning decision in such a way that we can very early realize whether they are actually fulfilled or not—including a fair understanding of what are significant deviations both in time and in scale—we cannot plan; and we have no feedback, no way of self-control in management.

We obviously also need for long-range planning *managerial* knowledge—the knowledge with respect to the operations of a business. We need such knowledge as that of the resources available, especially the human resources, their capacities and their limitations. We need to know how to "translate" from business needs, business results, and business decisions into functional capacity and specialized effort. There is, after all, no functional decision, there is not even functional data, just as there is no functional profit, no functional loss, no functional investment, no functional risk, no functional customer, no functional product, and no functional image of a company. There is only a unified company product, risk, investment and so on, hence only company performance and company results. Yet at the same time the work obviously has to be done by people each of whom has to be specialized. Hence for a decision to be possible, we must be able to integrate divergent individual knowledges and capacities into one organization potential; and for a decision to be effective, we must be able to translate it into a diversity of individual and expert, yet focused, efforts.

There are also big problems of knowledge in the entrepreneurial task that I have not mentioned—the problems of growth and change, for instance, or those of the moral values of a society and their meaning to business. But these are problems that exist for many areas and disciplines other than management.

And in this paper I have confined myself intentionally to knowledge that is specific to the process of long-range planning. Even so I have barely mentioned the main areas. But I think I have said enough to substantiate three conclusions:

(a) Here are areas of genuine knowledge, not just areas in which we need data. What we need above all, are basic theory and conceptual thinking.

(b) The knowledge we need is new knowledge. It is not to be found in the traditional disciplines of business such as accounting or economics. It is also not available, by and large, in the physical or life sciences. From the existing disciplines we can get a great deal of help, of course, especially in tools and techniques. And we need all we can get. But the knowledge we need is distinct and specific. It pertains not to the physical, the biological, or the psychological universe, though it partakes of them all. It pertains to the specific institution, the enterprise, which is a social institution existing in contemplation of human values. What is "knowledge" in respect to this institution, let alone what is "scientific," must, therefore, always be determined by reference to the nature, function, and purposes of this specific (and very peculiar) institution.

(c) It is not within the decision of the entrepreneur whether he wants to make risk-taking decisions with long futurity; he makes them by definition. All that is within his power is to decide whether he wants to make them responsibly or irresponsibly, with a rational chance of effectiveness and success, or as blind gamble against all odds. And both because the process is essentially a rational process, and because the effectiveness of the entrepreneurial decisions depends on the understanding and voluntary efforts of others, the process will be the more responsible and the more likely to be effective, the more it is a rational, organized process based on knowledge.

Long-range planning is risk-taking decision making. As such it is the responsibility of the policy maker, whether we call him entrepreneur or manager. To do the job rationally and systematically does not change this. Long-range planning does not "substitute facts for judgment," does not "substitute science for the manager." It does not even lessen the importance and role of managerial ability, courage, experience, intuition, or even hunch—just as scientific biology and systematic medicine have not lessened the importance of these qualities in the individual physician. On the contrary, the systematic organization of the planning job and the supply of knowledge to it should make more effective individual managerial qualities of personality and vision.

But at the same time, long-range planning offers major opportunity and major challenge to Management Science and to the Management Scientist.* We need systematic study of the process itself and of every one of its elements. We need systematic work in a number of big areas of new knowledge—at least we need to know enough to organize our ignorance.

At the same time, long-range planning is the crucial area; it deals with the decisions which, in the last analysis, determine the character and the survival of the enterprise.

So far, it must be said, Management Science has not made much contribution to long-range planning. Sometimes one wonders whether those who call themselves Management Scientists are even aware of the risk-taking character of economic activity and of the resultant entrepreneurial job of long-range planning. Yet, in the long run, Management Science and Management Scientists may well, and justly, be judged by their ability to supply the knowledge and thinking needed to make long-range planning possible, simple, and effective.

* I would like to say here that I do not believe that the world is divided into "managers" and "management scientists." One man may well be both. Certainly, management scientists must understand the work and job of the manager, and vice versa. But conceptually and as a kind of work the two are distinct.

9 | Business Objectives
and Survival Needs

The literature of business management, confined to a few "how to do" books only fifty years ago, has grown beyond any one man's capacity even to catalogue it. Professional education for business has become the largest and most rapidly growing field of professional education in this country and is growing rapidly in all other countries in the free world. It also has created in the advanced continuing education for experienced, mature, and successful executives—perhaps first undertaken in systematic form at the University of Chicago—the only really new educational concept in a hundred and fifty years.

Yet so far we have little in the way of a "discipline" of business enterprise, little in the way of an organized, systematic body of knowledge, with its own theory, its own concepts, and its own methodology of hypothesis, analysis, and verification.

First published in *The Journal of Business of the University of Chicago,* April 1958.

The Need for a Theory of Business Behavior

The absence of an adequate theory of business enterprise is not just an academic concern; on the contrary, it underlies four major problems central to business as well as to a free-enterprise society.

(1) One is the obvious inability of the layman to understand modern business enterprise and its behavior. What goes on, and why, "at the top" or "on the fourteenth floor" of the large corporation—the central economic and one of the central social institutions of modern industrial society—is as much of a mystery to the outsider as the magician's sleight of hand is to the small boy in the audience. And the outsiders include not only those truly outside business enterprise. They include workers and shareholders; they include many professionally trained men in the business—the engineers or chemists, for instance—indeed, they include a good many management people themselves: supervisors, junior executives, functional managers. They may accept what top management does but they accept on faith rather than by reason of knowledge and understanding. Yet such understanding is needed for the success of the individual business as well as for the survival of industrial society and of the free-enterprise system.

One of the real threats is the all-but-universal resistance to profit in such a system, the all-but-universal (but totally fallacious) belief that socialism—or any other ism—can operate an industrial economy without the rake-off of profit, and the all-but-universal concern lest profit be too high. That the danger in a dynamic, industrial economy is that profit may be too low to permit the risks of innovation, growth, and expansion—that, indeed, there may be no such thing as profit but only provision for the costs of the future —very few people understand.

This ignorance has resisted all attempt at education; this resistance to profits has proved impervious to all propaganda or appeals, even to the attempts at profit-sharing.

The only thing capable of creating understanding of the essential and necessary function of profit in an expanding, risk-taking, industrial economy is an understanding of business enterprise. And that, for all without personal, immediate experience in the general management of a business, can come only through a general "model" of business enterprise, that is, through the general theory of a systematic discipline.

(2) The second problem is the lack of any bridge of understanding between the macroeconomics of an economy and the microeconomics of the most important actor in this economy, the business enterprise. The only microeconomic concept to be found in economic theory today is that of "profit maximization." To make it fit the actual, observable behavior of business enterprise, however, economists have had to bend, stretch, and qualify it until it has lost all meaning and all usefulness. It has become as complicated as the "epicycles" with which pre-Copernican astronomers tried to save the geocentric view of the universe: profit maximization may mean short-run, immediate revenue or long-range basic profitability of wealth-producing resources; it may have to be qualified by a host of unpredictables such as managerial power drives, union pressures, technology, etc.; and it completely fails even then to account for business behavior in a growing economy. It does not enable the economist to predict business reaction to public policy any more; to the governmental policy maker, business reaction is as irrational as government policy, by and large, seems to the businessman.

But in modern industrial society we must be able to "translate" easily from public policy to business behavior and back again. The policy maker must be able to assess the impact of public policy on business behavior; and the businessman—especially in the large enterprise—must be able to assess the impact of his decisions and actions on the macro-economy. Profit-maximization does not enable us to do either, primarily because it fails to understand the role and function of profit.

(3) The third area in which the absence of a genuine theory of business enterprise creates very real problems is that of the internal integration of the organization. The management literature is full of discussions of the "problem of the specialist" who sees only his own functional area or of the "problem of the scientist in business" who resents the demand that he subordinate his knowledge to business ends. Yet we will be getting ever more specialized; we will, of necessity, employ more and more highly trained professionals. Each of those must be dedicated to his speciality; yet each must share a common vision and common goals and must voluntarily engage in a common effort. To bring this about is already the most time- and energy-consuming job of management, certainly in our big businesses, and no one I know claims to be able to do it successfully.

Twenty years ago it was still possible to see a business as a mechanical assemblage of "functions." Today we know that, when we talk of a business, the functions simply do not exist. There is only business profit, business risk, business product, business investment, and business customer. The functions are irrelevant to any one of them. And yet it is equally obvious, if we look at the business, that the work has to be done by people who specialize, because nobody can know enough even to know all there is to be known about one of the major functions today—they are growing too fast. It is already asking a great deal of a good man to be a good functional man, and, in some areas, it is rapidly becoming almost too much to ask of a man. How, then, do we transmute functional knowledge and functional contribution into general direction and general results? The ability of big business—but even of many small ones—to survive depends on our ability to solve this problem.

(4) The final problem—also a symptom both of the lack of discipline and of the need for it—is of course the businessman's own attitude toward theory. When he says, "This is theoretical," he by and large still means: "This is irrelevant." Whether managing a business enterprise could or should be a science (and one's answer to this question depends primarily on how one defines the word science), we need to be able to consider theory the foundation for good practice. We would have no modern doctors, unless medicine (without itself being a science in any strict sense of the word) considered the life-sciences and their theories the foundation of good practice. Without such a foundation in a discipline of business enterprise, we cannot make valid general statements, cannot, therefore, predict the outcome of actions or decisions, and can judge them only by hindsight and by their results—when it is too late to do anything. All we can have at the time of decision would be hunches, hopes, and opinions, and, considering the dependence of modern society on business enterprise and the impact of managerial decisions, this is not good enough.

Without such a discipline we could also neither teach nor learn, let alone work systematically on the improvement of our knowledge and of our performance as managers of a business. Yet the need both for managers and for constant improvement of their knowledge and performance is so tremendous, quantitatively as well as qualitatively, that we simply cannot depend on the "natural selection" of a handful of geniuses.

The need for a systematic discipline of business enterprise is particularly pressing in the underdeveloped growth countries of the world. Their ability to develop themselves will depend, above all, on their ability rapidly to develop men capable of managing business enterprise, that is, on the availability of a discipline that can be taught and can be learned. If all that is available to them is development through experience, they will almost inevitably be pushed toward some form of collectivism. For, however wasteful all collectivism is of economic resources, however destructive it is of freedom, dignity, and happiness, it economizes the managerial resource through its concentration of entrepreneurial and managerial decisions in the hands of a few planners at the top.

What Are the Survival Needs of Business Enterprise?

We are still a long way from a genuine "discipline" of business enterprise. But there is emerging today a foundation of knowledge and understanding. It is being created in some of our large companies and in some of our universities. In some places the starting point is economics, in some marketing, in some the administrative process, in others such new methodologies as operations or systems research or long-range planning. But what all these approaches, regardless of starting point or terminology, have in common is that they start out with the question: What are the survival needs of business enterprise? What, in other words, does it have to be, to do, to achieve—to exist at all? For each of these needs there has, then, to be an objective.

It may be said that this approach goes back to the pioneering work on business objectives that was done at the Bell Telephone System under the presidency of Theodore Vail a full forty years ago. Certainly, that was the first time the management of a large business enterprise refused to accept the old, glib statement, "The objective of a business is to make a profit," and asked instead, "On what will our survival as a privately owned business depend?" The practical effectiveness of the seemingly so obvious and simple approach is proved by the survival, unique in developed countries, of privately owned telecommunications in the United States and Canada. A main reason for this was certainly the "survival objective" Vail set for the Bell ·System: "Public satisfaction with our service." Yet, though proved in practice, this remained, until recently, an isolated example. And it probably had to remain such until, within the last generation, the biologists developed the approach to understanding of systems by means of defining "essential survival functions."

"Survival objectives" are general; they must be the same in general for each and every business. Yet they are also specific; different performance and different results would be needed in each objective area for any particular business. And every individual business will also need its own specific balance between them at any given time.

The concept of survival objectives thus fulfills the first requirement of a genuine theory—that it be both formal and yet concretely applicable, that is, practical. Survival objectives are also objective both as to their nature and as to the specific requirements in a given situation. They do not depend on opinion or hunch. Yet—and this is essential—they do not "determine" entrepreneurial or managerial decisions; they are not (as is so much of traditional economics or of contemporary behavioral science) an attempt to substitute formulas for risk-taking decision or responsible judgment. They attempt rather to establish the foundation for decision and judgment, to make what is the specific task of entrepreneur and manager possible, effective, and rational, and to make it understandable and understood.

We have reached the stage where we know the "functions" of a business enterprise, with function being used the way the biologist talks about procreation as a function essential for the perpetuation of a living species.

There are *five such survival functions* of business enterprise. Together they define the areas in which each business, to survive, has to reach a standard of performance and produce results above a minimum level. They are also the areas affected by every business decision and, in turn, affecting every business result. Together these five areas of survival objectives describe therefore (operationally) the nature of business enterprise.

(1) The enterprise needs, first, a *human organization designed for joint performance* and capable of perpetuating itself.

It is an assemblage not of brick and mortar but of people. These people must work as individuals; they cannot work any other way. Yet they must voluntarily work for a common result and must, therefore, be organized for joint performance. The first requirement of business is, therefore, that there be an effective human organization.

But business must also be capable of perpetuating itself as a human organization if only because all the things we decide every day—if, indeed, we are managers—take for their operation more time than the Good Lord has allotted us. We are not making a single decision the end of which we are likely to see while still working. How many managerial decisions will be liquidated within twenty years, will have disappeared, unless they are totally foolish decisions? Most of the decisions we make take five years before they even begin to have an impact; this is the short range of a decision. And then they take ten or fifteen years before (at the very earliest) they are liquidated, have ceased to be effective, and, therefore, have ceased to have to be reasonably right.

This means that the enterprise as a human organization has to be able to perpetuate itself. It has to be able to survive the life-span of any one man.

(2) The second survival objective arises from the fact that the enterprise exists in *society and economy*. In business schools and business thinking we often tend to assume that the business enterprise exists by itself in a vacuum. We look at it from the inside. But the business enterprise is a creature

of society and economy. If there is one thing we do know, it is that society and/or economy can put any business out of existence overnight—nothing is simpler. The enterprise exists on sufferance and exists only as long as society and economy believe that it does a job and a necessary, a useful, and a productive one.

I am not talking here of public relations; they are only one means. I am not talking of something that concerns only the giants. And I am not talking of socialism. Even if the free-enterprise system survives, individual businesses and industries within it may be—and of course often have been—restricted, penalized, or even put out of business very fast by social or political action such as taxes or zoning laws, municipal ordinances or federal regulation, and so forth. Anticipation of social climate and economic policy, on the one hand, and organized behavior to create what business needs to survive in respect to both are, therefore, genuine survival needs of each business at all times. They have to be considered in every action and have to be factored into every business decision.

Equally, the business is a creature of the economy and at the mercy of changes in it—in population and income, ways of life and spending patterns, expectations and values. Again here is need for objectives which anticipate so as to enable the business to adapt and which at the same time aim at creating the most favorable conditions.

(3) Then, of course, there is the area of the specific purpose of business, of its contribution. The purpose is certainly to *supply an economic good and service*. This is the only reason why business exists. We would not suffer this complicated, difficult, and controversial institution except for the fact that we have not found any better way of supplying economic goods and services productively, economically, and efficiently. So, as far as we know, no better way exists. But that is its only justification, its only purpose.

(4) There is another purpose characteristic which I would, so to speak, call the nature of the beast; namely, that this all happens in a *changing* economy and a *changing* technology. Indeed, in the business enterprise we have the first institution which is designed to produce change. All human institutions since the dawn of prehistory or earlier had always been designed to prevent change—all of them: family, government, church, army. Change has always been a catastrophic threat to human security. But in the business enterprise we have an institution that is designed to create change. This is a very novel thing. Incidentally, it is one of the basic reasons for the complexity and difficulty of the institution.

This means not only that business must be able to adapt to change—that would be nothing very new. It means that every business, to survive, must strive to *innovate*. And innovation, that is, purposeful, organized action to bring about the new, is as important in the social field—the ways, methods, and organization of business, its marketing and market, its financial and personnel management, and so on—as it is in the technological areas of product and process.

In this country industrial research expenditures have risen from a scant one-tenth of 1 percent of national income to 1½ or 2 percent in less than thirty years. The bulk of this increase has come in the last ten years; this means that the impact in the form of major technological changes is still ahead of us. The speed of change in nontechnological innovation, for instance, in distribution channels, has been equally great. Yet many businesses are still not even geared to adaptation to change; and only a mere handful are geared to innovation—and then primarily in the technological areas. Here lies, therefore, a great need for a valid theory of business enterprise but also a great opportunity for contribution.

(5) Finally, there is an absolute requirement of survival, namely, that of *profitability*, for the very simple reason that everything I have said so far spells out *risk*. Everything I have said so far says that it is the purpose, the nature, and the necessity of this institution to take risks, to create risks. *And risks are genuine costs.* They are as genuine a cost as any the accountant can put his finger on. The only difference is that, until the future has become past, we do not know how big a cost; but they *are* costs. Unless we provide for costs, we are going to destroy capital. Unless we provide for loss, which is another way of saying for future cost, we are going to destroy wealth. Unless we provide for risk, we are going to destroy capacity to produce. And, therefore, a minimum profitability, adequate to the risks which we, by necessity, assume and create, is an absolute condition of survival not only for the enterprise but for society.

This says three things. First, the need for profitability is objective. It is of the nature of business enterprise and as such is independent of the motives of the businessman or of the structure of the "system." If we had archangels running businesses (who, by definition, are deeply disinterested in the profit motive), they would have to make a profit and would have to watch profitability just as eagerly, just as assiduously, just as faithfully, just as responsibly, as the most greedy wheeler-dealer or as the most convincedly Marxist commissar in Russia.

Second, profit is not the "entrepreneur's share" and the "reward" to one "factor of production." It does not rank on a par with the other "shares," such as that of labor, for instance, but above them. It is not a claim *against* the enterprise but the claim *of* the enterprise—without which it cannot survive. How the profits are distributed and to whom is of great political importance; but for the understanding of the needs and behavior of a business it is largely irrelevant.

Finally, "profit maximization" is the wrong concept,

whether it be interpreted to mean short-range or long-range profits or a balance of the two. The relevant question is, "What minimum does the business need?"—not "What maximum can it make?" This "survival minimum" will, incidentally, be found to exceed present maxima in many cases. This, at least, has been my experience in most companies where a conscious attempt to think through the risks of the business has been attempted.

Here are five dimensions, and each of these five is a genuine view of the whole business enterprise. It is a human organization, and we can look upon it only in that aspect, as does our human relations literature. We can look at it from its existence in society and economy, which is what the economist does. This is a perfectly valid, but it is a one-sided view.

We can, similarly, look at the enterprise only from the point of view of its goods and services. Innovation and change are yet another dimension, and profitability is yet another. These are all genuine true aspects of the same being. But only if we have all five of them in front of us do we have a theory of business enterprise on which practice can be built.

For managing a business enterprise means making decisions, every one of which both depends on needs and opportunities in each of these five areas and, in turn, affects performance and results in each.

The Work to Be Done

The first conclusion from this is that every business needs objectives—explicit or not—in each of these five areas, for malfunction in any one of these endangers the entire business. And failure in any one area destroys the entire business

—no matter how well it does in the other four areas. Yet these are not interdependent but autonomous areas.

(1) Here, then, is the first task of a discipline of business enterprise: *to develop clear concepts and usable measurements to set objectives and to measure performance in each of these five areas.*

The job is certainly a big one—and a long one. There is no area as yet where we can really define the objectives, let alone measure results. Even in respect to profitability we have, despite great recent advances in managerial economics, figures for the past rather than measurements that relate current or expected profitability to the specific future risks and needs. In the other areas we do not even have that, by and large. And in some—the effectiveness of the human organization, the public standing in economy and society, or the area of innovation—we may, for a long time to come, perhaps forever, have to be content with qualitative appraisal making possible judgment. Even this would be tremendous progress.

(2) A second conclusion is hardly less important: *no one simple objective is "the" objective of a business; no one single yardstick "the" measure of performances, prospects, and results of a business; no one single area "the" most important area.*

Indeed, the most dangerous oversimplification of business enterprise may well be that of the "one yardstick," whether return on investment, market standing, product leadership, or what have you. At their best these measure performance in one genuine survival area. But malfunction or failure in any one area is not counterbalanced by performance in any other area, just as a sturdy respiratory or circulatory system will not save an animal if its digestive or nervous system collapses. Success, like failure, in business enterprise is *multidimensional.*

(3) This, however, brings out another important need: a rational and systematic approach to the *selection and bal-*

ance among objectives so as best to provide for survival and growth of the enterprise. These can be called the "ethics" of business enterprise, insofar as ethics is the discipline that deals with rational value choices among means to ends. It can also be the "strategy" of entrepreneurship. Neither ethics nor strategy is capable of being absolutely determined, yet neither can be absolutely arbitrary. We need a discipline here that encompasses both the "typical" decision which adapts to circumstances and "plays" the averages of statistical probability, and the innovating, "unique event" of entrepreneurial vision and courage, breaking with precedent and trends and creating new ones—and there are already some first beginnings of such a discipline of entrepreneurship. But such a discipline can never be more than theory of composition is to the musical composer or theory of strategy to the military leader: a safeguard against oversight, an appraisal of risks, and, above all, a stimulant to independence and innovation.

Almost by definition, the demands of different survival objectives pull in different directions, at least for any one time period. And it is axiomatic that the resources even of the wealthiest business, or even of the richest country, never cover in full all demands in all areas; there is never so much that there has to be no allocation. Higher profitability can thus be achieved only by taking a risk in market standing, in product leadership, or in tomorrow's human organization, and vice versa. Which of these risks the enterprise can take, which it cannot take, and which it cannot afford not to take— these risk-taking, value-decisions between goals in one area versus goals in others, and between goals in one area today versus goals in others tomorrow, is a specific job of the entrepreneur. This decision itself will remain a "judgment," that is, a matter of human values, appraisal of the situation, weighing of alternatives, and balancing of risks. But an understanding of survival objectives and their requirements can supply both the rational foundation for the decision itself and the rational criteria for the analysis and appraisal of entrepreneurial performance.

An Operational View of the Budgeting Process

The final conclusion is that we need a new approach to the process in which we make our value decisions between different objective areas—the budgeting process. And in particular do we need a real understanding of that part of the budget that deals with the expenses that express these decisions, that is, the "managed" and "capital" expenditures.

Commonly today, budgeting is conceived as a financial process. But it is only the notation that is financial; the decisions are entrepreneurial. Commonly today, managed expenditures and capital expenditures are considered quite separate. But the distinction is an accounting (and tax) fiction and misleading; both commit scarce resources to an uncertain future; both are, economically speaking, capital expenditures. And they, too, have to express the same basic decisions on survival objectives to be viable. Finally, today, most of our attention in the operating budget is given, as a rule, to other than the managed expenses, especially to the variable expenses, for that is where, historically, most money was spent. But, no matter how large or small the sums, it is in our decisions on the managed expenses that we decide on the future of the enterprise.

Indeed, we have little control over what the accountant calls variable expenses—the expenses which relate directly to units of production and are fixed by a certain way of doing things. We can change them, but not fast. We can change a relationship between units of production and labor costs (which we, with a certain irony, still consider variable expenses despite the fringe benefits). But within any time period these expenses can only be kept at a norm and cannot be changed. This is, of course, even more true for the expenses in respect to the decisions of the past, our fixed expenses. We cannot make them undone at all, whether these are capital expenses or taxes or what have you. They are beyond our control.

In the middle, however, are the expenses for the future which express our risk-taking value choices: the capital expenses and the managed expenses. Here are the expenses on facilities and equipment, on research and merchandising, on product development and people development, on management and organization. This managed expense budget is the area in which we really make our decisions on our objectives. (That, incidentally, is why I dislike accounting ratios in that area so very much, because they try to substitute the history of the dead past for the making of the prosperous future.)

We make decisions in this process in two respects. First, what do we allocate people for? For the money in the budget is really people. What do we allocate people, and energy, and efforts to? To what objectives? We have to make choices, as we cannot do everything.

And, second, what is the time scale? How do we, in other words, *balance* expenditures for long-term permanent efforts against any decision with immediate impact? The one shows results only in the remote future, if at all. The development of people (a fifteen-year job), the effectiveness of which is untested and unmeasurable, is, for instance, a decision on faith over the long range. The other may show results immediately. To slight the one, however, might, in the long range, debilitate the business and weaken it. And, yet, there are certain real short-term needs that have to be met in the business—in the present as well as in the future.

Until we develop a clear understanding of basic survival objectives and some yardsticks for the decisions and choices in each area, budgeting will not become a rational exercise of responsible judgment; it will retain some of the hunch character that it now has. But our experience has shown that the concept of survival objectives alone can greatly improve both the quality and effectiveness of the process and the understanding of what is being decided. Indeed, it gives us, we are learning, an effective tool for the integration of functional work and specialized efforts and especially for creating a common understanding throughout the organization and common measurements of contribution and performance.

The approach to a discipline of business enterprise through an analysis of survival objectives is still a very new and a very crude one. Yet it is already proving itself a unifying concept, simply because it is the first *general* theory of the business enterprise we have had so far. It is not yet a very refined, a very elegant, let alone a very *precise*, theory. Any physicist or mathematician would say: This is not a theory; this is still only rhetoric. But at least, while maybe only in rhetoric, we are talking about something real. For the first time we are no longer in the situation in which theory is irrelevant, if not an impediment, and in which practice has to be untheoretical, which means cannot be taught, cannot be learned, and cannot be conveyed, as one can only convey the general.

This should thus be one of the breakthrough areas; and twenty years hence this might well have become the *central* concept around which we can organize the mixture of knowledge, ignorance, and experience, of prejudices, insights, and skills, which we call "management" today.

10 | The Manager and the Moron

The computers, despite all the excitement they have been generating, are not yet economically important. It's only now that IBM is shipping them out at a rate of a thousand a month that they're even beginning to have an impact. But we haven't begun to use the potential of the computer. So far we are using it only for clerical chores, which are unimportant by definition. To be sure, the computer has created something that had never existed in the history of the world—namely, paying jobs for mathematicians. But that is hardly a major economic contribution, no matter what the graduate dean thinks.

So the economic impact of the new technologies is still in the future. If we subtracted every single one of them from the civilian economy, we would hardly notice it in the figures—perhaps a percentage point or two.

But this situation of linear movement is rapidly changing in every respect. And the greatest change is one that an economist, looking only at the figures, wouldn't even notice: In the past twenty years we have created a brand-new form of capital, a brand-new resource, namely, knowledge.

First published in *The McKinsey Quarterly,* Spring 1967.

Up until 1900, any society in the world would have done just as well as it did without men of knowledge. We may have needed lawyers to defend criminals and doctors to write death certificates, but the criminals would have done almost as well without the lawyers, and the patients without the doctors. We needed teachers to teach other ornaments of society, but this, too, was largely decoration. The world prided itself on men of knowledge, but it didn't need them to keep the society running.

As late as the mid-forties, General Motors carefully concealed the fact that one of its three top men, Albert Bradley, had a Ph.D. It was even concealed that he had gone to college, because, quite obviously, a respectable man went to work as a water boy at age fourteen. A Ph.D. was an embarrassing thing to have around.

Nowadays, companies boast about the Ph.D.'s on their payrolls. Knowledge has become our capital resource, a terribly expensive one. A man who graduates from a good business school represents some $100,000 of social investment, not counting what his parents spent on him, and not counting the opportunity costs. His grandparents and great-grandparents had to go to work at the age of twelve or thirteen with the hoe in the potato patch so that he could forgo those ten years of contribution to society. And that's a tremendous capital investment.

Besides spending all that money, we are also doing something very revolutionary. We are applying knowledge to work. Seven-odd thousand years ago, the first great human revolution took place when our ancestors first applied skill to work. They did not use skill to substitute for brawn. The most skilled work very often requires the greatest physical strength; no ditchdigger works harder than the surgeon performing a major operation. Rather, our ancestors put skills on top of physical labor. And now—a second revolution—we've put knowledge on top of both. Not as a substitute for skill, but as a whole new dimension. Skill alone won't do it any more.

Now, this has two or three important implications for management.

First, we must learn to make knowledge productive. As yet we don't really know how. The payroll cost of knowledge workers already amounts to more than half the labor costs of practically all businesses I know. That represents a tremendous capital investment in human beings. But so far neither productivity trends nor profit margins show much sign of responding to it. Pretty clearly, although business is paying for knowledge workers, it isn't getting much back. And if you look at the way we manage knowledge workers, the reason is obvious: we don't know how.

One of the few things we do know is that for any knowledge worker, even for the file clerk, there are two laws. The first one is that knowledge evaporates unless it's used and augmented. Skill goes to sleep, it becomes rusty, but it can be restored and refurbished very quickly. That's not true of knowledge. If knowledge isn't challenged to grow, it disappears fast. It's infinitely more perishable than any other resource we have ever had. The second law is that the only motivation for knowledge is achievement. Anybody who has ever had a great success is motivated from then on. It's a taste one never loses. So we do know a little about how to make knowledge productive.

The Obsolescence of Experience

Another implication flows from the creation of this new knowledge resource. The new generation of managers, those now aged thirty-five or under, is the first generation that thinks in terms of putting knowledge to work before one has accumulated a decade or two of experience. Mine was the last generation of managers who measured their value entirely by experience. All of us, of necessity, managed by experience—not a good process, because experience cannot be tested or be taught. Experience must be experienced; except by a very great artist, it cannot be conveyed.

This means that the new generation and my generation are going to be horribly frustrated working together. They rightly expect us, their elders and betters, to practice some of the things that we preach. We don't dream of it. We preach knowledge and system and order, since we never had them. But we go by experience, the one thing we do have. We feel frustrated and lost because, after devoting half our lifetimes to acquiring experience, we still don't really understand what we're trying to do. The young are always in the right, because time is on their side. And that means *we* have to change.

This brings us to the third implication, a very important one. Any business that wants to stay ahead will have to put very young people into very big jobs—and fast. Older men cannot do these jobs—not because they lack the necessary intelligence, but because they have the wrong conditioned reflexes. The young ones stay in school so long they don't have time to acquire the experience we used to consider indispensable for big jobs. And the age structure of our population is such that in the next twenty years, like it or not, we are going to have to promote people we wouldn't have thought old enough, a few years ago, to find their way to the water cooler. Companies must learn to stop replacing the sixty-five-year-old man with the fifty-nine-year-old. They must seek out their good thirty-five-year-olds.

For all its importance, however, the appearance of knowledge as a new capital resource is not the most vivid change in our environment, if only because it does not yet have a visible impact on the world's economic figures. Probably the most vivid change is in technology.

Many of the old technologies, of course, still have a lot of life in them. I think it's quite clear that the automobile, for instance, has yet to experience its greatest growth period. In the developed countries, however, it's in a defensive position. I don't think we need a great deal of imagination to foresee the day when the private car will be banned in the midtown areas or the day when the internal combustion engine will be limited to over-the-road use.

Or consider steel. I think one can quite easily foretell technological changes that will cut the cost of steel by about 40 percent. But whether that's enough to recreate momentum for the steel industry is debatable. I think that steel would probably need a greater cost advantage to make it again the universal material it used to be. Since steel, like all multipurpose materials, isn't ideal for any one use, it has to compete on price. And, as you know, the steel industry has lost 20 percent of the markets it had before World War II. It's concrete here, plastic there, and so on. Whether steel will lose the automotive body business to one of the new composition materials in the next ten years is a moot question. Only a fool would bet on it at this point, but by the same token only a fool would bet against it. If it does happen, it's very doubtful whether even a 40 percent reduction in cost might be enough to keep steel from joining the long parade of yesterday's engines of economic growth.

In agriculture, the great need is for an advance in productivity—but again, not in the developed countries. By now, the agricultural population in the developed countries has shrunk to such a small percentage of the total that even tripling its productivity would make little difference in the over-all economic picture.

And so on. I'm not saying that the industries based on old technologies can't advance, but I am saying they're unlikely to provide the impetus we need for continuing expansion. From now on, I think, the expansion will have to be powered by new industries based on new technologies, something we have not seen to any extent since before World War I.

Enter the Knowledge Utility

One of the most potentially earth-shaking forces in our economy is the technology of information. I don't mean simply the computer. The computer is to information what the electric power station is to electricity. The power station makes many other things possible, but it's not where the money is. The money is in the gimmicks and gizmos, the appliances, the motors and facilities made possible and necessary by electricity, that didn't exist before.

Information, like electricity, is energy. Just as electrical energy is energy for mechanical tasks, information is energy for mental tasks. The computer is the central power station, but there are also the electronic-transmission facilities—the satellites and related devices. We have devices to translate the energy, to convert the information. We have the display capacity of the television tube, the capability to translate arithmetic into geometry, to convert from binary numbers into curves. We can go from computer core to memory display, and from either one into hard copy. All the pieces of the information system are here. Technically, there is no reason why Sears, Roebuck could not offer tomorrow, for the price of a television set, a plug-in appliance that would put us in direct contact with all the information needed for schoolwork from kindergarten through college.

Already the time-sharing principle has begun to take hold. I don't think it takes too much imagination to see that a typical large company is about as likely to have its own computer twenty years hence as it is to have its own steam-generating plant today. It is reasonably predictable that computers will become a common carrier, a public utility, and that only organizations with quite extraordinary needs will have their own. Steel mills today have their own generators because they need such an enormous amount of power. Twenty years hence, an institution that's the equivalent of a steel mill in terms of mental work—MIT, for example—might well have its own computer. But I think most other universities, for most purposes, will simply plug into time-sharing systems.

It would be silly to try to predict in detail the effects of any development as big as this. All one can foresee for certain is a great change in the situation. One cannot predict what it will lead to, and where and when and how. A change as tremendous as this doesn't just satisfy existing wants, or replace things we are now doing. It creates new wants and makes new things possible.

A New Age of Information

The impact of information, however, should be greater than that of electricity, for a very simple reason. Before electricity, we had power; we had energy. It was very expensive and rather scarce, but we had it. Before now, however, we have not had information. Information has been unbelievably expensive, almost totally unreliable, and always so late that it was of little if any value. Most of us who had to work with information in the past, therefore, knew we had to invent our own. One developed, if one had any sense, a reasonably good instinct for what invention was plausible and likely to fly, and what wasn't. But real information just wasn't to be had. Now, for the first time, it's beginning to be available—and the over-all impact on society is bound to be very great.

Without attempting to predict the precise nature and

timing of this impact, I think we ca[n]
assumptions.

ASSUMPTION ONE: Within the next ten years, info[r]
become very much cheaper. An hour of computer ti[m]
costs several hundred dollars at a minimum; I have
figures that put the cost at about a dollar an hour in 1973
so. Maybe it won't come down that steeply, but come down
it will.

ASSUMPTION TWO: The present imbalance between the
capacity to compute and store information and the capacity
to use it will be remedied. We will spend more and more
money on producing the things that make a computer usable
—the software, the programs, the terminals, and so on. The
customers aren't going to be content just to have the computer sitting there.

ASSUMPTION THREE: The kindergarten stage is over. We're
past the time when everybody was terribly impressed by the
computer's ability to do two plus two in fractions of a
nanosecond. We're also past the stage of trying to find work
for the computer by putting all the unimportant things on
it—using it as a very expensive clerk. Actually, nobody has
yet saved a penny that way, as far as I can tell. Clerical
work—unless it's a tremendous job, such as addressing seven-
million copies of *Life* magazine every week—is not really
done very cheaply on the computer. But then, kindergartens
are never cheap.

Now we can begin to use the computer for the things it
should be used for—information, control of manufacturing
processes, control of inventory, shipments, and deliveries.
I'm not saying we shouldn't be using the computer for
payrolls, but that's beside the point. If payrolls were all it
could do, we wouldn't be interested in it.

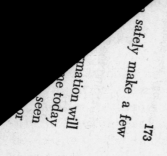

...at the computer makes no ...rs. It's a total moron, and ...es us to think, to set the ...e brighter the master has to ...we have ever had. All it can ...it can do that awfully fast. It ...charge overtime. It extends our capacity more th... ...l we have had for a long time, because of all the really unskilled jobs it can do. By taking over these jobs, it allows us—in fact, it compels us—to think through what we are doing.

But though it can't make decisions, the computer will—if we use it intelligently—increase the availability of information. And that will radically change the organization structure of business—of all institutions, in fact. Up to now we have been organizing, not according to the logic of the work to be done, but according to the absence of information. Whole organization levels have existed simply to provide standby transmission facilities for the breakdowns in information flow that one could always take for granted. Now these redundancies are no longer needed. We mustn't allow organizational structure to be made more complicated by the computer. If the computer doesn't enable us to simplify our organizations, it's being abused.

Along with vastly increasing the availability of information, the computer will reduce the sheer volume of data that managers have had to cope with. At present the computer is the greatest possible obstacle to management information, because everybody has been using it to produce tons of paper. Now, psychology tells us that the one sure way to shut off all perception is to flood the senses with stimuli.

That's why the manager with reams of computer output on his desk is hopelessly uninformed. That's why it's so important to exploit the computer's ability to give us *only* the information we want—nothing else. The question we must ask is not, "How many figures can I get?" but "What figures do I need? In what form? When and how?" We must refuse to look at anything else. We no longer have to take figures that mean nothing to us and read them the way a gypsy reads tea leaves.

Instead, we must decide on our information needs and how the computer can fill those needs. To do that, we must understand our operating processes, and the principles behind the processes. We must apply knowledge and analysis to them, and convert them to a clerk's routine. Even a work of genius, thought through and systematized, becomes a routine. Once it has been created, a shipping clerk can do it—or a computer can do it. So, once we have achieved real understanding of what we are doing, we can define our needs and program the computer to fill them.

Beyond the Numbers Barrier

We must realize, however, that we cannot put in the computer what we cannot quantify. And we cannot quantify what we cannot define. Many of the important things, the subjective things, are in this category. To *know* something, to really understand something important, one must look at it from sixteen different angles. People are perceptually slow, and there is no shortcut to understanding; it takes a great deal of time. Managers today cannot take the time to understand, because they don't have it. They are too busy working on things they can quantify—things they *could* put in a computer.

This is why the manager should use the computer to

control the routines of business, so that he himself can spend ten minutes a day controlling instead of five hours. Then he can use the rest of his time to think about the important things he cannot really know—people and environment. These are things he cannot define; he has to take the time to go and look. The failure to go out and look is what accounts for most of our managerial mistakes today.

Our greatest managerial failure rate comes in the step from middle to top management. Most middle managers are doing essentially the same things they did on their entrance jobs: controlling operations and fighting fires. In contrast, the top manager's primary function is to think. The criteria for success at the top level bear little resemblance to the criteria for promotion from middle management.

The new top manager, typically, has been promoted on the basis of his ability to adapt successfully. But suddenly he's so far away from the firing line that he doesn't know what to adapt to—so he fails. He may be an able man, but nothing in his work experience has prepared him to think. He hasn't the foggiest notion how one goes about making entrepreneurial or policy decisions. That's why the failure rate at the senior-management level is so high. In my experience, two out of three men promoted to top management don't make it; they stay middle management. They aren't necessarily fired. Instead, they get put on the Executive Committee with a bigger office, a bigger title, a bigger salary—and a higher nuisance value because they have had no exposure to thinking. This is a situation we are going to eliminate.

On the other hand, we are going to open up a new problem of development at the middle-management level. It isn't difficult for us to get people into middle management today. But it is going to be, because we shall need thinking people in the middle, not just at the top. The point at which we teach people to think will have to be moved further and further down the line. We can already see this problem in the big commercial banks.

We will have to manage knowledge correctly in order to preserve it. And this gets us into myriad questions of teaching and learning, of developing knowledge and techniques of thinking—not only in the developed nations, but in countries that are yet unaware of the distinction between management-by-experience and management-by-thinking, countries that are unaware of management itself. But that is another subject.

11 | The Technological Revolution: Notes on the Relationship of Technology, Science, and Culture

The standard answer to the question, "What brought about the explosive change in the human condition these last two hundred years?" is "The Progress of Science." This paper enters a demurrer. It argues that the right answer is more likely: "A fundamental change in the concept of technology." Central to this was the re-ordering of old technologies into systematic public disciplines with their own conceptual equipment, e.g., the "differential diagnosis" of nineteenth-century medicine. In the century between 1750 and 1850 the three main technologies of Man—Agriculture, the Mechanical Arts (today's Engineering), and Medicine—went in rapid succession through this process, which resulted almost immediately in an agricultural, an industrial, and a medical "revolution" respectively.

This process owed little or nothing to the new knowledge of contemporary science. In fact, in every technology the practice with its rules of thumb was far ahead of science. Technology, therefore, became the spur to science; it took, for instance, seventy-five years until Clausius and Kelvin could give a scientific formulation to the thermodynamic

First published in *Technology and Culture*, Fall 1961.

behavior of Watt's steam engine. Science could, indeed, have had no impact on the Technological Revolution until the transformation from craft to technological discipline had first been completed.

But technology had an immediate impact on science, which was transformed by the emergence of systematic technology. The change was the most fundamental one—a change in science's own definition and image of itself. From being "natural philosophy," science became a social institution. The words in which science defined itself remained unchanged: "the systematic search for rational knowledge." But "knowledge" changed its meaning from being "understanding," i.e., focused on man's mind, to being "control," i.e., focused on application in and through technology. Instead of raising, as science had always done, fundamental problems of metaphysics, it came to raise, as it rarely had before, fundamental social and political problems.

It would be claiming too much to say that technology established itself as the paramount power over science. But it was technology that built the future home, took out the marriage license, and hurried a rather reluctant science through the ceremony. And it is technology that gives the union of the two its character; it is a coupling of science *to* technology, rather than a coupling of science and technology.

The evidence indicates that the key to this change lies in new basic concepts regarding technology, that is, in a genuine Technological Revolution with its own causes and its own dynamics.

Of all major technologies medicine alone has been taught systematically for any length of time. An unbroken line leads back for one thousand years, from the medical school of today to the medical schools of the Arab caliphates. The trail, though partly overgrown, goes back, another fourteen hundred years, through the School of Alexandria to Hippocrates. From the beginning, medical schools taught both theoretical knowledge and clinical practice, engaged simultaneously, therefore, in science and technology. Unlike any other technologist in the West, the medical practitioner has continuously enjoyed social esteem and position.

Yet, until very late—1850 or thereabouts—there was no organized or predictable relationship between scientific knowledge and medical practice. The one major contribution to health care which the West made in the Middle Ages was the invention of spectacles. The generally accepted date is 1286; by 1290 the use of eyeglasses is fully documented.* This invention was, almost certainly, based directly upon brand-new scientific knowledge, most probably on Roger Bacon's optical experiments. Yet Bacon was still alive when spectacles came in—he died in 1294. Until the nineteenth century there is no other example of such all but instantaneous translation of new scientific knowledge into technology—least of all in medicine. Yet Galen's theory of vision, which ruled out any mechanical correction, was taught in the medical schools until 1700.†

Four hundred years later, in the age of Galileo, medicine took another big step—Harvey's discovery of the circulation of the blood, the first major new knowledge since the ancients. Another hundred years, and Jenner's smallpox vaccination brought both the first specific treatment and the first prevention of a major disease.

* E. Rosen, "The Invention of Eyeglasses," *Journal for the History of Medicine*, 11 (1956), pp. 13–46, 183–218.

† It was among the great Boerhaave's many "firsts" to have taught the first course in opthalmology and to examine actual eyes—in 1708 in Leyden. Newton's *Optics* was the acknowledged inspiration. (See George Sarton, "The History of Medicine versus the History of Art," *Bulletin of the History of Medicine*, 10 [1941], pp. 123–35.)

Harvey's findings disproved every single one of the theoretical assumptions that underlay the old clinical practice of bleeding. By 1700 Harvey's findings were taught in every medical school and repeated in every medical text. Yet bleeding remained the core of medical practice and a universal panacea for another hundred years, and was still applied liberally around 1850.* What killed it finally was not scientific knowledge—available and accepted for two hundred years—but clinical observation.

In contrast to Harvey, Jenner's achievement was essentially technological and without any basis in theory. It is perhaps the greatest feat of clinical observation. Smallpox vaccination had hard sledding—it was, after all, a foolhardy thing deliberately to give oneself the dreaded pox. But what no one seemed to pay any attention to was the complete incompatibility of Jenner's treatment with any biological or medical theory of the time, or of any time thereafter until Pasteur, one hundred years later. That no one, apparently, saw fit to try explaining vaccination or to study the phenomenon of immunity appears to us strange enough. But how can one explain that the same doctors who practiced vaccination, for a century continued to teach theories which vaccination had rendered absurd?

The only explanation is that science and technology were not seen as having anything to do with one another. To us it is commonplace that scientific knowledge is being translated into technology, and vice versa. This assumption explains the violence of the arguments regarding the historical relationship between science and the "useful arts." But the assumptions of the debate are invalid: the presence of a tie

---

* Bleeding actually reached a peak in the 1820's, when it was touted as the universal remedy by no less an authority than Broussais, the most famous professor at the Paris Academy of Medicine. According to Henry E. Sigrist (*The Great Doctors,* New York, 1933), it became so popular that in the one year, 1827, 33 million leeches were imported into France.

proves as little as its absence—it is our age, not the past, which presumes consistency between theory and practice.

The basic difference was not in the content but in the focus of the two areas. Science was a branch of philosophy, concerned with understanding. Its object was to elevate the human mind. It was misuse and degradation of science to use it—Plato's famous argument. Technology, on the other hand, was focused on use. Its object was increase of the human capacity to do. Science dealt with the most general, technologies with the most concrete. Any resemblances between the two were "purely coincidental."*

There are no hard and fast dates for a major change in an attitude, a world view. And the Technological Revolution was nothing less. We do know, however, that it occurred

* There was, to be sure, one famous dissent, one important and highly effective approach to science as a means to doing and as a foundation for technology. Its greatest spokesman was St. Bonaventura, the thirteenth-century antiphonist to St. Thomas Aquinas (see especially St. Bonaventura's *Reduction of all Arts to Theology*). A hundred years earlier the dissenters actually dominated in the twelfth-century Platonism of the theologian–technologist schools of St. Victoire and Chartres, builders alike of mysticism and of the great cathedrals. On this see Charles Homer Haskin, *The Renaissance of the twelfth century* (Cambridge, Mass., 1927); Otto von Simpson *The Gothic Cathedral* (New York, 1956); and *Abbot Suger and the Cathedral Church of St. Denis*, edited by Erwin Panofsky (Princeton, 1946).

The dissenters did not, of course, see material technology as the end of knowledge; rational knowledge was a means towards the knowledge of God or at least His glorification. But knowledge, once its purpose was application, immediately focused on material technology and purely worldly ends—as St. Bernard pointed out in his famous attack on Suger's "technocracy" as early as 1127.

The dissent never died down completely. But after the Aristotelian triumph of the thirteenth century, it did not again become respectable, let alone dominant until the advent of Romantic Natural Philosophy in the early nineteenth century, well *after* the Technological Revolution and actually its first (and so far only) literary offspring. It is well known that there was the closest connection between the Romantics—with Novalis their greatest poet, and with Schelling their official philosopher—and the first major discipline which, from its inception, was always both science and technology: organic chemistry. Less well known is the fact that the Romantic movement, its philosophers, writers, and statesmen came largely out of the first technical university, the Mining Academy in Freiberg (Saxony) that had been founded in 1776.

within the half century 1720 to 1770—the half century that separates Newton from Benjamin Franklin.

Few people today realize that Swift's famous encomium on the man who makes two blades grow where one grew before, was not in praise of the scientist. On the contrary, it was the final, crushing argument in a biting attack on them, and especially against the august Royal Society. It was meant to extol the sanity and benefits of nonscientific technology against the arrogant sterility of an idle enquiry into nature concerned with understanding; this is against Newtonian Science, for Swift was, as always, on the unpopular side. But his basic assumption—that science and application were radically different and worlds apart—was clearly the prevailing one in the opening decades of the eighteenth century. No one scientist spoke out against the weirdest technological "projects" of the South Sea Bubble of 1720, even though their theoretical infeasibility must have been obvious to them. Many, Sir Isaac Newton taking the lead, invested heavily in them.* And while Newton, as Master of the Royal Mint, reformed its business practices, he did not much bother with its technology.

Fifty years later, around 1770, Dr. Franklin is the "philosopher" par excellence and the West's scientific lion. Franklin, though a first-rate scientist, owed his fame to his achievements as a technologist—"artisan" in eighteenth-century parlance. He was a brilliant gadgeteer, as witness the Franklin stove and bifocals. Of his major scientific exploits, one—the investigation of atmospheric electricity—was immediately turned into useful application: the lightning rod. Another, his pioneering work in oceanography with its discovery of the Gulf Stream, was undertaken for the express purpose of application, viz., to speed up the transatlantic mail service. Yet the scientists hailed Franklin as enthusiastically as did the general public.

In the fifty years between 1720 and 1770—not a particu-

* J. Carswell, *The South Sea Bubble* (Stanford, 1959).

larly distinguished period in the history of science, by the way—a fundamental change in the attitude toward technology, both of laity and of scientists, must have taken place. One indication is the change in English attitude towards patents. During the South Sea Bubble they were still unpopular and attacked as "monopolies." They were still given to political favorites rather than to an inventor. By 1775 when Watt obtained his patent, they had become the accepted means of encouraging and rewarding technological progress.

We know in detail what happened to technology in the period which includes both the Agricultural Revolution and the opening of the Industrial Revolution. Technology as we know it today, that is, systematic, organized work on the material tools of man, was born then. It was produced by collecting and organizing existing knowledge, by applying it systematically, and by publishing it. Of these steps the last one was both the most novel—craft skill was not for nothing called a "mystery"—and the most important.

The immediate effect of the emergence of technology was not only rapid technological progress: it was the establishment of technologies as systematic disciplines to be taught and learned and, finally, the reorientation of science towards feeding these new disciplines of technological application.

Agriculture* and the mechanical arts† changed at the same time, though independently.

* G. E. Fussell, *The Farmer's Tools, 1500–1900* (London, 1952); A. J. Bourde, *The Influence of England on the French Agronomes* (Cambridge, 1953); A. Demolon, *L'Evolution Scientifique et l'Agriculture Francaise* (Paris, 1946); R. Krzymowski; *Geschichte der deutscen Landwirtschaft* (Stuttgart, 1939).

† A. P. Usher, *History of Mechanical Inventions* (Rev. Ed., Cambridge, Mass., 1954); also the same author's "Machines & Mechanisms" in Volume III of Singer, *et al.*, *A History of Technology* (Oxford, 1957); J. W. Roe, *English and American Tool Builders* (New Haven, 1916); K. R. Gilbreth; "Machine Tools," in *A History of Technology*, Vol. IV (Oxford, 1958); on early technical education see: Franz Schnabel, *Die Anfaenge des Technischen Hochschulwesens* (Freiburg, 1925).

Beginning with such men as Jethro Tull and his systematic work on horse-drawn cultivating machines in the early years of the seventeenth century and culminating towards its end in Coke of Holkham's work on balanced large-scale farming and selective livestock breeding, agriculture changed from a "way of life" into an industry. Yet this work would have had little impact but for the systematic publication of the new approach, especially by Arthur Young. This assured both rapid adoption and continuing further work. As a result, yields doubled while manpower needs were cut in half—which alone made possible that large-scale shift of labor from the land into the city and from producing food to consuming food on which the Industrial Revolution depended.

Around 1780, Albrecht Thaer in Germany, an enthusiastic follower of the English, founded the first agricultural college—a college not of "farming" but of "agriculture." This in turn, still in Thaer's lifetime, produced the first, specifically application-focused new knowledge, namely, Liebig's work on the nutrition of plants, and the first science-based industry, fertilizer.

The conversion of the mechanical arts into a technology followed the same sequence and a similar time table. The hundred years between the 1714 offer of the famous £20,000 prize for a reliable chronometer and Eli Whitney's standardization of parts was, of course, the great age of mechanical invention—of the machine tools, of the prime movers, and of industrial organization. Technical training, though not yet in systematic form, began with the founding of the *École des Ponts et Chaussées* in 1747. Codification and publication in organized form goes back to Diderot's *Encyclopédie*, the first volume of which appeared in 1750. In 1776—that miracle year that brought the Declaration of Independence, *The Wealth of Nations*, Blackstone's *Com-*

mentaries, and Watt's first practical steam engine—the first modern technical university opened: the *Bergakademie* (Mining Academy) in Freiberg, Saxony. Significantly enough, one of the reasons for its establishment was the need for technically trained managers created by the increasing use of the Newcomen steam engine, especially in deep-level coal mining.

In 1794, with the establishment of the *École Polytechnique* in Paris, the profession of engineer was established. And again, within a generation, we see a reorientation of the physical sciences—organic chemistry and electricity begin their scientific career, being simultaneously sciences and technologies. Liebig, Woehler, Faraday, Henry, Maxwell were great scientists whose work was quickly applied by great inventors, designers, and industrial developers.

Only medicine, of the major technologies, did not make the transition in the eighteenth century. The attempt was made—by the Dutchman Gerhard van Swieten,* not only a great physician but politically powerful as advisor to the Hapsburg Court. Van Swieten attempted to marry the clinical practice which his teacher Boerhaave had started at Leyden around 1700 with the new scientific methods of such men as the Paduan Morgagni whose *Pathological Anatomy*† (1761) first treated diseases as afflictions of an organ rather than as "humors." But—a lesson one should not forget—the very fact that medicine (or rather, something by that name) was already respectable and organized as an academic faculty defeated the attempt. Vienna relapsed into

---

* The standard biography of van Swieten is W. Mueller, *Gerhard van Swieten* (Vienna, 1883); on the organized resistance of academic medicine to the scientific approach, see G. Strakosch-Grassmann, *Geschichte des oesterreichischen Unterrichtswesens* (Vienna, 1905).

† This is the name commonly used for the work. Its actual title was *De Sedibus et causis morborum per anatomen indigatis;* the first English translation appeared in 1769 under the title, *The Seats and Causes of Diseases Investigated by Anatomy.*

medical scholasticism as soon as van Swieten and his backer, the Emperor Joseph II, died.

It was only after the French Revolution had abolished all medical schools and medical societies that a real change could be effected. Then another court physician, Corvisart, Napoleon's doctor, accomplished, in Paris around 1820, what van Swieten had failed in. Even then, opposition to the scientific approach remained powerful enough to drive Semmelweis out of Vienna and into exile when he found, around 1840, that traditional medical practices were responsible for lying-in fever with its ghastly death toll. Not until 1850, with the emergence of the modern medical school in Paris, Vienna, and Wuerzburg, did medicine become a genuine technology and an organized discipline.

This, too, happened, however, without benefit of science. What was codified and organized was primarily old knowledge, acquired in practice. Immediately *after* the reorientation of the practice of medicine, the great medical scientists appeared—Claude Bernard, Pasteur, Lister, Koch. And they were all application-focused, all driven by a desire to do, rather than by a desire to know.

We know the results of the Technological Revolution, and its impacts. We know that, contrary to Malthus, food supply in the last two hundred years has risen a good deal more than an exploding human population. We know that the average life-span of man a hundred fifty years ago was still close to the "natural life-span": the twenty-five years or so needed for the physical reproduction of the species. In the most highly developed and prosperous areas, it has almost tripled. And we know the transformation of our lives through the mechanical technologies, their potential, and their dangers.

Most of us also know that the Technological Revolution has resulted in something even more unprecedented: a common world civilization. It is corroding and dissolving history, tradition, culture, and values throughout the world, no matter how old, how highly developed, how deeply cherished and loved.

And underlying this is a change in the meaning and nature of knowledge and of our attitude to it. Perhaps one way of saying this is that the non-Western world does not want Western science primarily because it wants better understanding. It wants Western science because it wants technology and its fruits. It wants control, not understanding. The story of Japan's Westernization between 1867 and her emergence as a modern nation in the Chinese War of 1894 is the classical, as it is the earliest, example.*

But this means that the Technological Revolution endowed technology with a power which none of the "useful arts"—whether agricultural, mechanical, or medical—had ever had before: impact on man's mind. Previously, the useful arts had to do only with how man lives and dies, how he works, plays, eats, and fights. How and what he thinks, how he sees the world and himself in it, his beliefs and values, lay elsewhere—in religion, in philosophy, in the arts, in science. To use technological means to affect these areas was traditionally "magic"—considered at least evil, if not asinine to boot.

With the Technological Revolution, however, application and cognition, matter and the mind, tool and purpose, knowledge and control have come together for better or worse.

There is only one thing we do not know about the Technological Revolution—but it is essential: What happened to bring about the basic change in attitudes beliefs, and values which released it? Scientific progress, I have tried to show,

* This is brought out most clearly in William Lockwood, *The Economic Development of Japan, 1868–1938* (Princeton, 1954).

had little to do with it. But how responsible was the great change in world outlook which, a century earlier, had brought about the great Scientific Revolution? What part did the rising capitalism play? And what was the part of the new, centralized national state with its mercantilistic policies on trade and industry and its bureaucratic obsession with written, systematic, rational procedures everywhere? (After all, the eighteenth century codified the laws as it codified the useful or applied arts.) Or do we have to do here with a process, the dynamics of which lie in technology? Is it the "progress of technology" which piled up to the point when it suddenly turned things upside down, so that the "control" which nature had always exercised over man now became, at least potentially, control which man exercises over nature?

This should be, I submit, a central question both for the general historian and for the historian of technology.

For the first, the Technological Revolution marks one of the great turning points—whether intellectually, politically, culturally, or economically. In all four areas the traditional —and always unsuccessful—drives of systems, powers, and religions for world domination is replaced by a new and highly successful world-imperialism, that of technology. Within a hundred years, it penetrates everywhere and puts, by 1900, the symbol of its sovereignty, the steam engine, even into the Dalai Lama's palace in Lhasa.

For the historian of technology, the Technological Revolution is not only the cataclysmic event within his chosen field; it is the point at which such a field as technology emerges. Up to that point there is, of course, a long and exciting history of crafts and tools, artifacts and mechanical ingenuity, slow, painful advances and sudden, rapid diffusion. But only the historian, endowed with hindsight, sees this as technology, and as belonging together. To contemporaries, these were separate things, each belonging to its own sphere, application, and way of life.

Neither the general historian nor the historian of technology has yet, however, concerned himself much with the Technological Revolution. The first—if he sees it at all—dismisses technology as the bastard child of science. The only general historian of the first rank (excepting only that keen connoisseur of techniques and tools, Herodotus) who devotes time and attention to technology, its role and impact is, to my knowledge, Franz Schnabel.* That Schnabel taught history at a technical university (Karlsruhe) may explain his interest. The historians of technology, for their part, tend to be historians of materials, tools, and techniques rather than historians of technology. The rare exceptions tend to be nontechnologists such as Lewis Mumford or Roger Burlingame who, understandably, are concerned more with the impact of technology on society and culture than with the development and dynamics of technology itself.

Yet technology is important today precisely because it unites both the universe of doing and that of knowing, connects both the intellectual and the natural histories of man. How it came thus to be in the center—when it always before had been scattered around the periphery—has yet to be probed, thought through, and reported.

* Franz Schnabel, *Deutsche Geschichte im 19. Jahrhundert* (4 vols., Freiburg, 1929–1937); the discussion of technology and medicine is found chiefly in Vol. III.

12 | Can Management Ever Be a Science?

Some time ago I was asked by one of the management associations to make a speech on "Management Science in Business Planning." I used this invitation to do something I had long intended to do, which was to scan the last four or five years of literature in the areas of management science: operations research; statistical theory and statistical decision making; systems theory, cybernetics, data processing, and information theory; econometrics, management accounting, and accounting theory; and so on. I also looked fairly closely at the management science work done in a number of businesses, either by their own staffs or by outside consultants.

No one, I am convinced, can read this literature or can survey the work done without being impressed by the potential and promise of management science. To be sure, managing will always remain somewhat of an art; the talent, experience, vision, courage, and character of the managers will always be major factors in their performance and in that of their enterprises. But this is true of medicine and doctors,

An address delivered at the Fiftieth Anniversary Conference of the Harvard Business School, September 1958.

too. And, as with medicine, management and managers—especially the most highly endowed and most highly accomplished managers—will become the more effective as their foundation of organized systematic knowledge and organized systematic search grows stronger, and as their roots in a real discipline of management and entrepreneurship grow deeper. That such a discipline is possible, the work already done in management science proves.

But no one, I am also convinced, can survey the work to date without being worried at the same time. The potential is there—but it is in danger of being frittered away. Instead of a management science which supplies knowledge, concepts, and discipline to manager and entrepreneur, we may be developing a management gadget bag of techniques for the efficiency expert.

The bulk of the work today concerns itself with the sharpening of already existing tools for specific technical functions—such as quality control or inventory control, warehouse location or freight-car allocation, machine loading, maintenance scheduling, or order handling. And, in fact, a good deal of the work is little more than a refinement of industrial engineering, cost accounting, or procedures analysis. Some, though not very much, attention is given to the analysis and improvement of functional efforts—primarily those of the manufacturing function but also, to some extent, of marketing and of money management.

But there is almost no work, no organized thought, no emphasis on managing an enterprise—on the risk-making, risk-taking, decision-making job. In fact, I could find only two examples of such work: the industrial dynamics program at Massachusetts Institute of Technology* and the operations research and synthesis work done in some parts of

* See Jay W. Forrester, "Industrial Dynamics: A Major Breakthrough for Decision Makers," HBR July–August 1958, p. 37.

the General Electric Company. Throughout management science—in the literature as well as in the work in progress —the emphasis is on techniques rather than on principles, on mechanics rather than on decisions, on tools rather than on results, and, above all, on efficiency of the part rather than on performance of the whole.

However, if there is one fundamental insight underlying all management science, it is that the business enterprise is a *system* of the highest order: a system the "parts" of which are human beings contributing voluntarily of their knowledge, skill, and dedication to a joint venture.* And one thing characterizes all genuine systems, whether they be mechanical like the control of a missile, biological like a tree, or social like the business enterprise: it is interdependence. The whole of a system is not necessarily improved if one particular function or part is improved or made more efficient. In fact, the system may well be damaged thereby, or even destroyed. In some cases the best way to strengthen the system may be to *weaken* a part—to make it *less* precise or *less* efficient. For what matters in any system is the performance of the whole; this is the result of growth and of dynamic balance, adjustment, and integration rather than of mere technical efficiency.

Primary emphasis on the efficiency of parts in management science is, therefore, bound to do damage. It is bound to optimize precision of the tool at the expense of the health and performance of the whole. (That the enterprise is a social rather than a mechanical system makes the danger all the greater, for the other parts do not stand still. They either respond so as to spread the maladjustment throughout the system or organize for sabotage.)

This is hardly a hypothetical danger. The literature abounds in actual examples—inventory controls that improve production runs and cut down working capital but fail

* See Kenneth E. Boulding, "General Systems Theory," *Management Science*, April 1956, p. 197.

to consider the delivery expectations of the customer and the market risks of the business; machine-loading schedules that overlook the impact of the operations of one department on the rest of the plant; forecasts that assume the company's competitors will just stand still; and so on.

Technically this is all excellent work. But therein lies its danger. The new tools are so much more powerful than the old tools of technical and functional work—the tools of trial and error and of cut and fit—that their wrong or careless use must do damage.

For management science to become a gadget bag, therefore, not only means a missed opportunity; it may also mean loss of its potential to contribute altogether, if not its degeneration into a mischief maker.

Hence the questions arise: Is it inevitable that management science become a gadget bag? Or would this be the result of something management science does today or fails to do? And what would be the requirements for a real management science that supplies the knowledge and the methodology we need?

The first clue lies, perhaps, in the origin of this new "management science" approach—and the origin is an unusual one indeed.

Every other discipline of man began with a crude attempt to define what its subject was. Then people set to work fashioning concepts and tools for its study. But management science began with the application of concepts and tools developed within a host of other disciplines for their own particular purposes. It may have started with the heady discovery that certain mathematical techniques, hitherto applied to the study of the physical universe, could also be applied to the study of business operations.

As a result, the focus of much of the work in management science has *not* been on such questions as: "What is the business enterprise? What is managing? What do the two do, and what do the two need?" Rather, the focus has been on: "Where can I apply my beautiful gimmick?" The emphasis has been on the hammer rather than on driving in the nail, let alone on building the house. In the literature of operations research, for instance, there are several dissertations along the lines of "155 applications of linear programming," but I have not seen any published study on "typical business opportunities and their characteristics."

What this indicates is a serious misunderstanding on the part of the management scientist of what "scientific" means. "Scientific" is not—as many management scientists naively seem to think—synonymous with quantification. If this were true, astrology would be the queen of the sciences. It is not even the application of the "scientific method." After all, astrologers observe phenomena, derive the generalization of a hypothesis therefrom, and then test the hypothesis by further organized observation. Yet astrology is superstition rather than science because of its childish assumption that there is a real zodiac, that the signs in it really exist, and that their fancied resemblance to some such earthly creature as a fish or a lion defines their character and properties (whereas all of them are nothing but the mnemonic devices of the navigators of antiquity).

In other words "scientific" presupposes a rational definition of the universe of the science (that is, of the phenomena which it considers to be real and meaningful) as well as the formulation of basic assumptions or postulates which are appropriate, consistent, and comprehensive. This job of defining the universe of a science and of setting its basic postulates has to be done, however crudely, *before* the scientific method can be applied. If it is not done, or done wrongly, the scientific method cannot be applied. If it is done, and done right, the scientific method becomes applicable and, indeed, powerful.

This idea is, of course, nothing new. It goes back to the distinction between the premises that are generally valid and those that pertain to a specific discipline, made in Aristotle's *Analytica Posteriora*. On the rediscovery of this principle during the last century rests the power of modern science and of its methods.*

Management science still has to do this job of defining its universe. If it does this, then all the work done so far will become fruitful—at least as preparation and training ground for real achievement. The first task for management science, if it is to be able to contribute rather than distort and mislead, is, therefore, to define the specific nature of its subject matter. This might include as a basic definition the insight that the business enterprise is a system made up of human beings. The assumptions, opinions, objectives, and even the errors of people (and especially of managers) are thus primary *facts* for the management scientist. Any effective work in management science really has to begin with analysis and study of them.

Starting, then, with this recognition of what there is to be studied, management science must next establish its basic assumptions and postulates—without which no science can develop proper methods. It might first include the vital fact that every business enterprise exists in economy and society; that even the mightiest is the servant of its environment by which it can be dismissed without ceremony, but that even the lowliest affects and molds the economy and society instead of just adapting to them; in other words, that the business enterprise exists only in an economic and social ecology of great complexity.

The basic postulates might include the following ideas:

* For a statement of the modern position, see Howard Eves and Carroll V. Newsom, *Foundations and Fundamental Concepts of Mathematics* (New York, Rinehart & Company, Inc., 1958), pp. 29–30.

(1) The business enterprise produces neither things nor ideas but humanly determined values. The most beautifully designed machine is still only so much scrap metal until it has utility for a customer.

(2) Measurements in the business enterprise are such complex, not to say metaphysical, symbols as money—at the same time both highly abstract and amazingly concrete.

(3) Economic activity, of necessity, is the commitment of present resources to an unknowable and uncertain future—a commitment, in other words, to expectations rather than to facts. Therefore, risk is of the essence, and risk making and risk taking constitute the basic function of enterprise. And risks are not only taken by the "general manager," but right through the whole organization by everybody who contributes knowledge—that is, by every manager and professional specialist. This risk is something quite different from risk in the statistician's probability; it is the risk of the unique event, the irreversible qualitative breaking of the pattern.

(4) Inside and outside the business enterprise there is constant irreversible change; indeed, the business enterprise exists as the agent of change in an industrial society, and it must be capable both of purposeful evolution to adapt to new conditions and of purposeful innovation to change the conditions.

Some of this is often said in the preface of books on management science. It generally stays in the preface, however. Yet for management science to contribute to business understanding, let alone become a science, postulates like the foregoing ought to be the fabric of its work. Of course we need quantification—though it tends to come fairly late in the development of a discipline (only now, for instance, can scientists really quantify in biology). We need the scientific method. And we need work on specific areas and operations—careful, meticulous detail work. But, above all, we need to recognize the particular character of business enterprise and the unique postulates necessary for its study. It is on this vision that we must build.

The first need of a management science is, then, that it respect itself sufficiently as a distinct and genuine discipline.

The second clue to what is lacking in management science as applied today is the emphasis throughout its literature and throughout its work on "minimizing risk" or even on "eliminating risk" as the goal and ultimate purpose of its work.

To try to eliminate risk in business enterprise is futile. Risk is inherent in the commitment of present resources to future expectations. Indeed, economic progress can be defined as the ability to take greater risks. The attempt to eliminate risks, even the attempt to minimize them, can only make them irrational and unbearable. It can only result in that greatest risk of all: rigidity.

The main goal of a management science must be to enable business to take the right risk. Indeed, it must be to enable business to take *greater* risks—by providing knowledge and understanding of alternative risks and alternative expectations; by identifying the resources and efforts needed for desired results and by mobilizing energies for the greatest contribution; and by measuring results against expectations, thereby providing means for early correction of wrong or inadequate decisions.

All this may sound like mere quibbling over terms. Yet the terminology of "risk minimization" does induce a decided animus against risk taking and risk making—that is, against business enterprise—in the literature of management science. Much of it echoes the tone of the technocrats of a generation ago, for it wants to subordinate business to technique, and it seems to see economic activity as a sphere of physical determination rather than as an affirmation and exercise of responsible freedom and decision.

This is worse than being wrong. This is lack of respect for one's subject matter—the one thing no science can afford and no scientist can survive. Even the best and most serious work of good and serious people—and there is no lack of them in management science—is bound to be vitiated by it.

The second requirement for a management science is, then, that it take its subject matter seriously.

There would be little reason for concern about the trend of management science if we did not need so badly a genuine discipline of entrepreneurship and business management.

We need a systematic supply of organized knowledge for the risk-making and risk-taking decisions of business enterprise in our complex and rapidly changing technology, economy, and society; tools for the measurement of expectations and results; effective means for common vision and communication among the many functional and professional specialists—each with his own knowledge, his own logic, and his own language—whose combined efforts are needed to make the right business decisions, to make them effective, and to produce results. We need something teachable and learnable if only because we need far too many people with managerial vision and competence to depend on the intuition of a few "natural-born" geniuses; and only the generalizations and concepts of a discipline can really be learned or taught.

We know that these are urgent needs. In fact, the future of the free enterprise system may depend on our ability to make major managerial and entrepreneurial decisions more rationally, and to make more people capable of making and of understanding such decisions.

There would be little reason for concern here if management science had not demonstrated its great potential to fill our need. Of course, it is only in its infancy; real knowledge and understanding in vitally important areas may be decades away—may, indeed, never be obtained. But the work already done is exciting and powerful, and the talent at work is of a high order of competence, ability, and dedication.

All this, however, may come to nought if management science permits itself to become a management gadget bag. The opportunity will be lost, the need will go unfulfilled, and the promise will be blighted unless management science learns to respect both itself and its subject.

Index